PROJECT CONTROLS: NEEDS and SOLUTIONS

Proceedings of a Specialty Conference sponsored
by the Construction Division of the
American Society of Civil Engineers

Lincolnshire Illinois Conference Center
Chicago, Illinois
June 8–9, 1987

Edited by C. William Ibbs and David B. Ashley

Published by the
American Society of Civil Engineers
345 East 47th Street
New York, New York 10017-2398

ABSTRACT

This volume contains the proceedings of an ASCE Specialty Conference on *Project Controls: Needs and Solutions*. The papers presented at this meeting and contained in this publication identify the most pressing problems with the current practice of project control and the systems, whether computerized or manual, that are employed to manage both design and construction efforts. These articles also propose and discuss possible solution strategies to these identified problems. For reasons of expediency, project control in this vein is limited to cost, schedule, and estimating issues. Special areas of emphasis include change management, design-construction system integration, conceptual cost control, performance assessment, extant control techniques, scope definition, and available new technologies. A concluding paper ties all the individual recommendations together, sorts and resolves the conflicts between these suggestions, and offers an implementation strategy and timetable.

Library of Congress Cataloging-in-Publication Data

Project controls.

Includes indexes.
1. Construction industry—Management—Data processing—Congresses. 2. Construction industry—Cost control—Congresses. I. Ibbs, C. William. II. Ashley, David B. III. American Society of Civil Engineers. Construction Division.
TH438.P755 1987 624'.068 87-11452
ISBN 0-87262-601-6

Library of Congress Catalog Card No.: 87-11452
ISBN 0-87262-601-6
Manufactured in the United States of America.

PREFACE

Many rapid and unforeseeable developments have transpired in the past decade in the field of construction project control. The affordability and dramatic growth in the use of computers and computerized systems are but one noteworthy example. Today more attention is dedicated to the quality of data collected and the transformation of that data into usable information than ever before. One may even assert that project control has evolved into a scientific, teachable set of skills rather than an experience-based art form.

To be sure, we still have grave problems with the ways we control our building projects. Changes and an appropriate mechanism to price them are serious problems. So are reliability of collected data, and integration of cost information with schedule data. And though people for years have been dreaming of truly integrated (a very overworked term) systems, we still remain some distance away from a design system—let alone a CAD system—that can "talk" transparently to a construction project control system.

The members of ASCE's Estimating, Cost and Schedule Control Committee have labored long and hard over resolution to some of these problems. We eventually decided that one way to draw some attention to these issues and generate good ideas was to sponsor an ASCE Specialty Conference. As it so happened, Jim Neil of Morrison-Knudsen and the Construction Industry Institute's Cost & Schedule Task Force had at about the same time initiated several research studies. They quickly and nicely evolved into a group of relevant papers for this conference. Thus, this Specialty Conference represents the first effort to unite the work of an ASCE committee and a CII Task Force. Hopefully, there will be many other happy experiences like this in the future.

The papers contained in this volume represent the hard work of all the authors and are most sincerely appreciated. Similarly, Ivan Beliveau, Jesus De La Garza, John Melin, Mike O'Connor and Randolph Thomas served as reviewers and are to be acknowledged for their efforts. These papers were subjected to rigorous peer review and are eligible for discussion in the *ASCE Journal of Construction Management and Engineering* and ASCE awards. The New York staff of ASCE, particularly Elizabeth Yee, Denise Nurse and Shiela Menaker, lent behind the scenes aid that outsiders can never fully appreciate. We do though.

This conference has not, of course, solved all our project control problems. It has, however, hopefully cast some light on the needs of users while identifying some feasible solution paths. ASCE's Estimating, Cost and Schedule Control committee will continue to work on these problems and always encourage qualified people to join and work with our group.

C. William Ibbs
University of Illinois
University of California,
Berkeley

David B. Ashley
University of Texas,
Austin

CONTENTS

MODEL PLANNING and CONTROLLING SYSTEM for ENGINEERING, PROCUREMENT and CONSTRUCTION of INDUSTRIAL PROJECTS

S. K. Nethery, Jr. *

INTRODUCTION

An industrial facility is built by an owner to support the strategic business plans of the company. The on-line need date of this facility is usually critical, and the time available between recognition of the need for it and the date that it is needed is usually less than that which will permit engineering, procurement, construction and startup phases to be accomplished in sequence. A solution is to overlap these phases. When this is done, the challenge still is to:

... Meet owner operational need dates.
... Meet all owner quality standards.
... Optimize cost and schedule without compromising on quality.
... Establish conditions for optimum productivity.
... Facilitate early detection of potential cost and schedule overruns.

The purpose of this paper is to describe a model planning and controlling system for engineering, procurement, construction and startup of an industrial project which meets this challenge. The system described is based on these key philosophies and conditions:

... The starting point for planning is the owner's need date for the facility and proceeds in reverse order fashion from this need date.
... Completeness and quality of detailed engineering work packages are such that they will permit construction contracting strategy to be handled on a fixed-price contract basis.
... Procurement activity planned to satisfy both engineering and construction need dates.
... Team atmosphere established among project participants that assures coordination over the project's life.
... Controlling system that effectively intergrates or interrelates the many component project functions and provides managers with timely information for forecasting and responsive control.

In presenting this model system the project phases and their included activities will be described first. Next, the details of the overall control structure. Finally, the control methods appropriate for each phase will be described.

* Project Control Manager, Dow Chemical U.S.A., Engineering & Construction Division, 400 West Belt South, Houston, Texas 77253-3387

PROJECT PHASES AND THEIR CHARACTERISTICS

The Phases

The life-cycle of an industrial facility encompasses the total time between identification of a requirement for the facility through its final demobilization. This paper is concerned with that portion of the life-cycle between the time an authorization is given to proceed with engineering and when the facility is operational. Within this time period there are five phases:

... Conceptual Engineering
... Detailed Engineering
... Procurement
... Construction
... Startup

When these phases are overlapped their timing relationship is as illustrated in Figure 1.

Characteristics of Conceptual Engineering Phase

Basic functional characteristics of the project are described during this phase and can be divided into:

... Prepartaion of process flow diagrams, heat and material balances, and equipment performance specifications.
... Perform studies to select the best operational conditions among available alternatives.
... Preparation of mechanical flow diagrams, piping line list, instrument summary sheets, electrical one-line diagrams, equipment arrangements, etc.
... Special instructions.

Quality execution of these provides a firm definition of project scope so as to minimize changes later during detailed engineering. As illustrated in Figure 1, the ability to influence final project cost is greatest in the conceptual engineering phase and rapidly diminishes thereafter.

A project strategy in the form of a "Project Execution Plan" is developed. This plan describes the basic organization and responsibilities of the project team members, contracting method(s) to be employed, labor availability, general timing, risk allocation, assessment of potential problems and contingency options, etc.

Other actions performed during this phase are:

... Initiate procurement on long-lead-time items.
... Initiate plans for obtaining required licenses and/or permits.
... Develop cost control accounts and breakdown structures to be used for project control. Cross-reference these to the owner's structure (if necessary).
... Prepare preliminary project budget.
... Establish initial master project schedule to determine the overall project duration and the interrelationships between project phases.

Constructability is an issue that must be considered from this phase through the construction phase. Since decisions made during this phase establish the physical arrangement for the project, constructability must be a test against which each conceptual decision is evaluated.

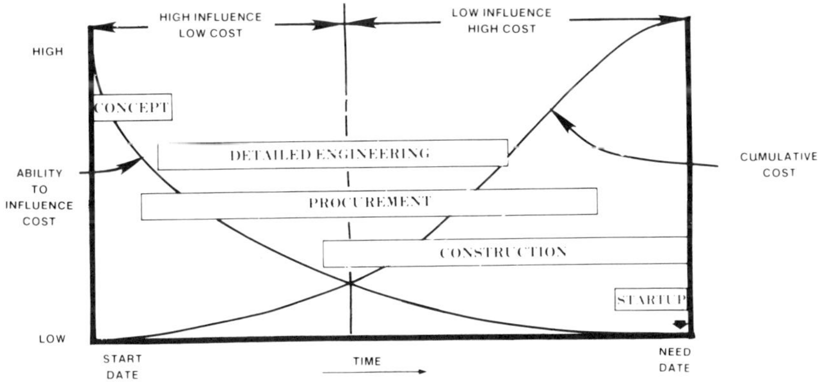

Figure 1. ABILITY TO INFLUENCE COST OVER PROJECT LIFE

<u>Characteristics of Detailed Engineering Phase</u>

The end products of this phase are work packages to support procurement and construction work. Detailed engineering activity is performed based on parameters defined during the conceptual engineering phase. These parameters are verified and expanded throughout this phase.

This phase is divided into:

... Preparation of technical procurement work packages.
... Preparation of construction work packages.

While construction work generally proceeds in bottom-up sequence, the detailed engineering phase will not exactly mirror this sequencing. It is generally from the top-down. This is a consequence of needing to have detailed information about the varied equipment, and how each item is connected together before detailed engineering can be done for foundations and structures.

Other actions are:

... Finalize the "Project Execution Plan".
... Refine detailed engineering schedule and expand construction schedule. Assure proper interface with procurement schedule.
... Establish the project control cost estimate and cash flow.
... Intensify constructability studies.
... Obtain licenses and/or permits.

<u>Characteristics of Procurement Phase</u>

This phase has two parallel and overlapping actions:

... Procurement of equipment and materials.
... Award of construction contracts.

With the procurement phase it is essential that:

... Equipment/material vendor information be available by the need dates designated on the detailed engineering schedule.

... The required equipment/material items be delivered by the need dates designated on the construction schedule.

... Contractors be moblized and ready to perform per start dates on the construction schedule.

Characteristics of Construcation Phase

In this phase the facility is constructed according to work packages prepared during the detailed engineering phase using equipment and material obtained in the procurement phase. The sequencing of construction will be initially planned to reflect the most logical and cost-effective approach to meeting startup need dates. As engineering and procurement schedules are further developed total compliance with this philosophy may not be possible and compromises will be required.

Characteristics of Startup Phase

The typical industrial facility is an assembly of multiple operating systems. Startup activity does not commence until all of the components that make up one of these operating systems are available for testing as a package. Consequently, construction activity relating to an operating system must be completed concurrently so that coordination will exist with needs of the startup schedule. Turnover to the owner occurs when an operating system is accepted by the owner. It can be accomplished system by system, or for the project as a whole.

PROJECT CONTROL STRUCTURE

Total control of a project involves individual control of each phase (Conceptual Engineering, Detailed Engineering, Procurement, Construction, and Startup) and their intergration.

The Schedule

The project schedule control milestone is the owner's need date. To meet this date requires that:

... Startup schedule reflect the owner's project need date.

... Construction schedule be coordinated with the startup schedule.

... Procurement schedule for equipment/materials take into consideration need dates associated with vendor information required for completion of engineering and need dates for delivery of equipment/materials to a designated site.

... Procurement schedule for construction take into consideration the time for qualification and selection of contractors.

... Detailed engineering schedule be coordinated with both the procurement schedule and construction schedule so that work packages are available when required.

... Conceptual engineering schedule be coordinated with the need dates for initation of procurement on long-lead-time items and the start of detailed engineering.

The initial schedule for a project is via a summary bar chart that displays key milestone dates. As definition of the project is improved, the schedule for each phase is described using a logic network to which the critical path method (CPM) is applied. CPM activities are at the level of detail that will permit effective progress reporting. Current work progress is frequently compared with the control schedule to determine if any problem has developed that will seriously affect meeting key project milestone dates.

The Budget

A preliminary project budget is prepared during the conceptual engineering phase using factor estimating techniques. The control budget is established from a definitive estimate prepared during the early stage of the detailed engineering phase. The estimate is updated as detailed engineering progresses, equipment/material purchase orders issued, and construction contracts awarded. Close attention is given to the variance between the control budget and the current estimate to determine if changes will have any dynamic impact on the project cost.

Breakdown Structure

There are two related control systems for a project. One system controls the work and is based upon the work breakdown structure (WBS); the other controls the cost and is based upon the cost breakdown structure (CWS). The WBS is composed of those direct labor elements of the project for which work-hours (WH) are budgeted and lend themselves to work progress measurement. The CBS is composed of all project elements for which dollars ($) are budgeted. The WBS elements are included within the CBS. Figure 2 illustrates this CBS/WBS relationship.

PHASES	INDIRECTS	DIRECTS			
		Labor		Material	Equipment
Conceptual Engineering	$	WH	$	$	$
Detail Engineering	$	WH	$	$	$
Procurement	$	WH	$	$	$
Construction	$	WH	$	$	$
Startup	$	WH	$	$	$
Other [1]	$				

Cost Breakdown Structure (CBS) — Work Breakdown Structure (WBS)

Footnotes 1 Home office overhead, contingency, profit, etc

Figure 2. TOTAL PROJECT COST BREAKDOWN STRUCTURE AND WORK BREAKDOWN STRUCTURE RELATIONSHIP

The WBS is used to build the project schedule control system and is based upon a database that is structured such that potential work tasks on any project are listed in generic fashion (standard code of accounts). This structure directly supports data collection of actual performance against schedule activities thus facilitating progress control, and can be summarized against control accounts for historical purposes and productivity studies. When a new project is planned this database becomes the reference for establishment of WBS activity values. Figure 3 displays the relationship between control accounts and schedule activities for the construction phase. Comparable control accounts/schedule activity relationships exist in the other project phases.

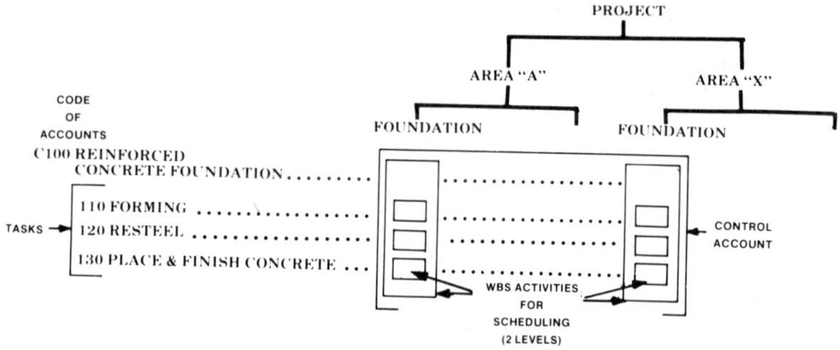

Figure 3. CONTROL ACCOUNT/SCHEDULE ACTIVITY RELATIONSHIP

A control account has these characteristics:

... A logical level for estimating, budgeting, and accumulating actual quantities and work-hours.

... May apply to one or more activities of the same type in the WBS.

... Hierarchical within the generic database.

Reporting

Work progress is reported at the tasks or activity level and analyzed using the earned value techniques. Cost reporting is normally done at the control account level. Since control accounts are utilized in defining the WBS and CBS a roll up of information to hierarchical levels is used to generate reports to satisfy various management groups.

PROJECT CONTROL WITHIN PHASES

General Approach

Planning and controlling of a phase requires:

... Dividing the phase into controllable segments (e.g. -- procurement trains, installation areas, operating systems).

... Defining the WBS activities within each segment.

... Estimating work-hours and/or cost value for each activity.
... Determining the duration for each activity.
... Identifying the relationships between activities within the segment, with related activities in other segments of the phase, with related activities in other phases, and with project milestones.

Having done the above, a logic network is prepared to which the critical path method (CPM) is applied.

Planning proceeds in reverse order -- from the startup phase back through the conceptual engineering phase -- using the owner's need date as the starting point. This is illustrated in Figure 4. The following discussion parallels this philosophy.

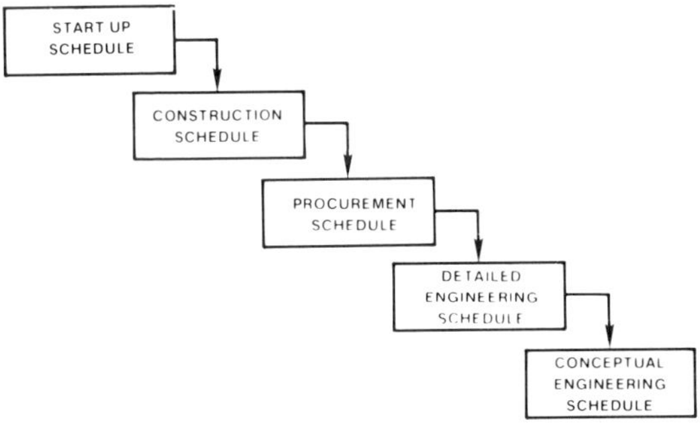

Figure 4. SCHEDULING SEQUENCE

Startup Control

A startup schedule that uses scoping of the facility by its individual operating systems permits control of this phase to handle the overlap sequencing at which these various systems must be brought on stream.

If the startup of any operating system will involve introduction of flammable feedstock, it is essential that special attention be given to any hot work (welding, stress relieving, etc.) that is to be completed in the facility after its introduction. It is preferable that completion of all hot work be done before introduction on these flammables.

Construction Control

In order for the construction schedule to reflect the way construction work is to be performed and also be in elements that can be controlled, the facility is divided into controllable geographic areas. Hard logic items (foundations, equipment setting, etc.) are very readily defined in the WBS for use in the CPM. Soft logic items (pipe erection, conduit installation, etc.) may be handled at a lesser level of detail since this type work at times does not follow strict logic.

Control for installation of individual pieces is handled using tabulations or bar charts displaying timing for each task based on related activity schedule dates from the construction schedule. Progress is reported at the task level and rolled up to the activity level. This procedure allows use of the earned value technique to determine percent complete for varied reporting levels.

As a matter of policy, work should not be started on any work package until applicable drawings and specifications are on hand, and acceptable delivery of related equipment/material is assured.

When construction progress has reached about 70-85% complete, the logic of remaining work activities is revised in a manner that it can be related to startup operating systems. This allows for coordination between this phase and the startup phase.

Procurement Control

Control of equipment/material procurement is by a schedule developed for individual procurement trains. Because so many agencies are involved in a procurement train the individual work activities that make up the inquiry-order-vendor documents-fabrication-deliver sequence are required elements for detailed control. A subsystem within the overall project control system is one for materials management that includes a warehousing module to track equipment/material between receipt at the designated site and their issue for use.

Dependent upon the contracting method used, the planning of construction procurement requires that adequate time be allowed for any of the applicable following:

... Development of a bidders list.
... Assembly of work package bidding documents (inquiry package).
... Issuance of the inquiry package to bidders.
... Preparation of bids by contractors.
... Pre-bid orientation and site visits.
... Extension of bidding time due to inquiry addenda.
... Analyzing the bids.
... Awarding the construction contract.
... Premobilization orientation meeting.
... Mobilization by contractor.

Detailed Engineering Control

Procurement and construction activities establish need dates for completion of various engineering actions. Thus, before the schedule for activities within the detailed engineering phase can be finished, it is necessary for the procurement phase to be defined in terms of controllable procurement trains and the construction into controllable areas of the facility.

The detailed engineering schedule must also reflect the way engineering deliverables are logically developed. Designing, calculating, drafting, and checking are separate activities that must be related both within and among engineering disciplines. Control for preparation of individual documents of a work package is handled using the same technique as for construction.

Scope control and change management must receive particular attention during detailed engineering since any scope growth or change can have a significant effect on remaining phases thus maybe affecting the ability to meet established project milestones.

Conceptual Engineering Control

Control of this phase is by a schedule based on the individual operating systems that make up the facility. The nature of this phase requires that initial planning use WBS activities that are broad in scope. As the scope of the basic functional features of the project are determined, activities can be expanded to a greater level of detail thus improving controllability.

General

Although detailed engineering and procurement are listed as separate phases, they are very interdependent in those instances where vendor documents of engineered items are required before continuation of detailed engineering.

For this system to succeed, performance of the detailed engineering phase must be intergrated with the needs of the procurement and construction phases. To do so often precludes efficient and/or cost-effective utilization of engineering personnel. This also means that the detailed engineering effort may cost more using this system than if completion of detailed engineering preceded the start of construction.

All cost/quantity estimates should parallel the work packaging for detailed engineering and construction. This approach facilitates estimate refinement as the project proceeds from the conceptual engineering phase to the detailed engineering phase.

The financial accounting/payroll function of the company must adapt to the project controls program, not vice-versa, since project control involves more detail and requires a more involved coding structure than does accounting. Also, the project control accounts must be responsive to the contract terms. These terms may establish a structure quite different from that used for normal company accounting and payroll. Of course, the accounting subsystem should link with the intergrated project control system so that most basic work-hours and cost data need only be recorded once to serve the systems.

The system described in this document is idealized. Real world conditions can be expected to interfere with the process at times. It is at these times that the "team" concept becomes so important because it will take an atmosphere of open communication and cooperation among the project team to brainstorm solutions, make compromises, and otherwise do what is necessary to maintain the momentum toward achievement of project objectives.

SUMMARY

A typical industrial facility is an intricate arrangement of structures, vessels, machinery, piping, electrical systems, and instrumentation packed into a relatively small area. The conceptual engineering, detailed engineering, procurement, construction and startup phases associated with such a facility would be complicated enough if each could be accomplished without dependence on the others. But, the fact is they are very interdependent and these dependences become increasingly critical as the phases are overlapped.

The model planning and controlling system described in this document is believed to be one that provided a system to determine optimum overlap -- the one which produces a project on schedule, within budget, and of the desired quality. It will shorten time as much as possible while still optimizing costs and preserving quality. The owner's need date is the ultimate milestone of the system with all contributing activities carefully planned to meet that date. No official claims of time or cost savings are made although users of this system feel comfortable with time savings in the order of 15% compared to the sequential approach, and construction labor savings as high as 30% as compared to fast-track approach which permits construction to be initiated on the basis of incomplete designs. The time savings accrue from the planned overlap. The work-hour savings accrue due to a planning and controlling system for a project which reduces work-hour requirements (constructability reviews), minimizes rework (complete work packages available at start of construction), assures materials availability when needed (procurement tracking and materials management), and stimulated productivity (effective project management and a team approach to everything).

Finally, it is appropriate to re-emphasize several major thoughts incorporated throughout the description of this system. First, the more attention given to a project during its definition, the greater is the potential for time and cost savings. Second, for the system to work, construction must yield to the needs of startup, engineering must yield to the needs of procurement and construction, and procurement must yield to the needs of construction. Third, there must be a continuous team approach to planning and execution which involves all project team groups.

REFERENCE:

Digest of "Model Planning and Controlling System for Engineering, Procurement and Construction of Industrial Projects", Publication, Construction Industry Institute, Cost/Schedule Task Force, 1987

PROJECT CONTROL TECHNIQUES

Leland. S. Riggs, M. ASCE*

Abstract: Most major construction companies use sophisticated, well-established systems for cost and schedule control. These systems, which have evolved over a number of years, rely on an extensive data base and many systems integrate cost and schedule while others make no such attempt. This paper summarizes the cost and schedule control techniques of several owners and contractors operating in the industrial sector.

Introduction

In November, 1982 the Business Roundtable published a report entitled "Modern Management Systems" [Business82]. This report noted that the construction industry has been slow to accept the latest management methods for project control. The report went on to state that many people view this lack of acceptance of modern management methods to be a primary cause of schedule delays and cost overruns.

In June, 1984 the Construction Industry Institute funded a research project to examine the cost and schedule control techniques which have been found to be effective in the industrial sector of construction. This paper is a summary of the findings of that research report [Riggs86].

The cost and schedule control techniques included in this paper were taken from briefings and interviews with three owner companies and four contractor companies primarily engaged in the industrial sector of construction. The contractor companies discussed in this paper are all listed in the top forty of the Engineering News Record 400 list of contractors. Likewise, the owner companies in this paper are equally successful in their fields.

Seven companies is an admittedly small sample, however these companies were selected because their techniques seemed to be advanced applications of project control and the techniques had proven to be successful. The point should be made, however, that this paper is not attempting to portray these techniques as representative of the industrial sector of the construction industry.

---

*Assoc. Prof., School of Civ. Engrg., Ga. Inst. of Tech., Atlanta, Ga.

Confidentiality

The research from which the data of this paper was taken was conducted at a level of confidentiality imposed by the Construction Industry Institute. Consequently, neither the owner nor the contractor companies are named in this paper.

Control Techniques

The techniques discussed in this paper are categorized under the following headings:

- Planning
- Scheduling
- Work Statusing
- Trend Analysis
- Forecasting

Planning Techniques

Work Breakdown Structure - All of the companies mentioned in this paper use some form of work breakdown. Fig. 1 is a portion of a typical work breakdown structure. The depth of levels on a breakdown can be any number to satisfy the requirements of the project. The number of levels observed for the companies included in this paper is between five and six.

One contractor defines the last level in a Work Breakdown Structure as a work package. The criteria for this definition was that the work could be done within a week. It is worth noting that other definitions of work packages exist in the lexicon. Neil defines work packages as having a well-defined scope, usually ending in a deliverable product, and identifiable in a numerical accounting system [Neil82]. Using this definition, a work package could end at any level shown in Fig. 1.

Also, for those integrated systems we observed there is a correspondence at every level between the cost control accounts and the work breakdown structure. A mechanism for tieing these two accounts together is the code of accounts. Fig. 2 is an explanation of the structure of the code of accounts used in Fig. 1. This is one of many possible systems. For example, another system might use additional digits beyond the work task to account for resources.

During our discussions with contractors, they noted instances where owners imposed an additional requirement for coding costs on a project. The owners needed to account for capitalized costs or for taxes. In any event, the requirement for essentially another set of books was invariably confusing and added to the contractor's work load. Several contractors pointed out that the addition of digits to their established code of accounts could capture most if not all the owners special needs.

Another instance of owner imposed accounts is the Federal Energy Regulatory Commission requirement that their coding structure be applied to every regulated power plant. This is in addition to any account the contractor might have established for project control.

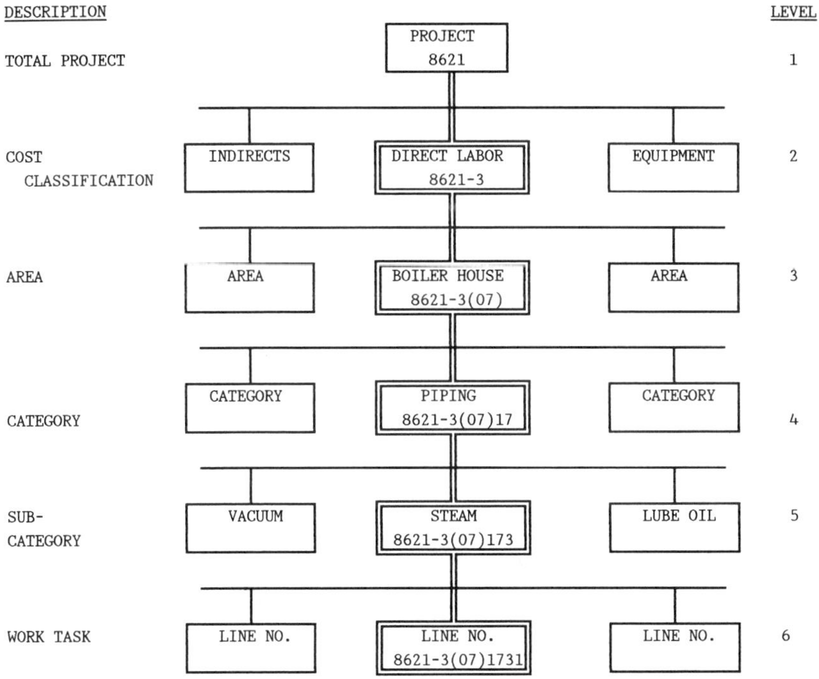

Fig. 1 Piping Work Breakdown Structure
Cost Codes adapted from [Thomas86]

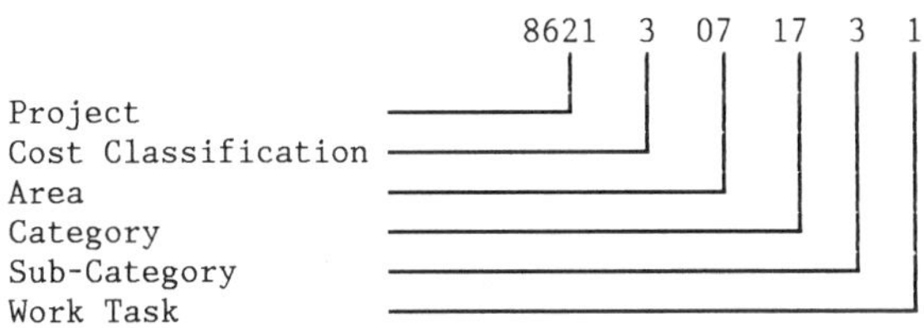

Fig. 2 Code of Accounts Structure

Schedule Techniques

Critical Path Method (CPM) - All the companies mentioned in this paper had the capability to do computerized CPM networks, however, the use of CPM seemed to be declining. For those companies who regularly use CPM, one effective technique was the use of generic schedules. These are schedules derived from years of historical records on similar projects. The generic schedule is retrieved from the computer and modified as necessary for the current project.

None of the companies reported using CPM below Level 4 on a Work Breakdown Structure. At Levels 5 and 6, the preference was decidedly for bar charts. At this level, contractors prefer to use 30 - 180 "look ahead" schedules and they find that bar charts better suit their purpose. Additionally, company representatives reported that CPM does not play as important a role as it did in the past. Many government projects, however, still require the use of CPM so the contractor companies at least retain a CPM capability.

The number of activities on CPM schedules varied greatly, from as low as 150 to several thousand. While some owners placed a requirement on contractors for a minimum number of activities, one owner held contractors to a maximum number of activities.

Bar Charts - Bar charts or Gantt charts are the technique of choice for scheduling and monitoring work in the field. Fig. 3 is a partial bar chart used by a contractor company. Bar charts of this type come from a Level 4 schedule.

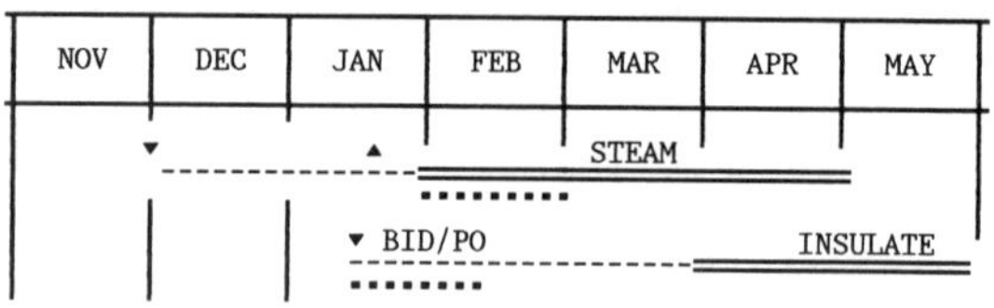

Fig. 3 - Contractor Bar Chart

Level 4 bar charts then become the basis for a weekly or Level 5 bar chart schedule. Fig. 4 is an example of a contractor's weekly detail schedule. In this instance, this contractor was working four ten-hour days.

Program Evaluation and Review Technique (PERT) and Line of Balance (LOB) are techniques which have been used in some segments of the construction industry. But apparently the owners and contractors studied in our research have found little need for these techniques and are therefore not included in this paper.

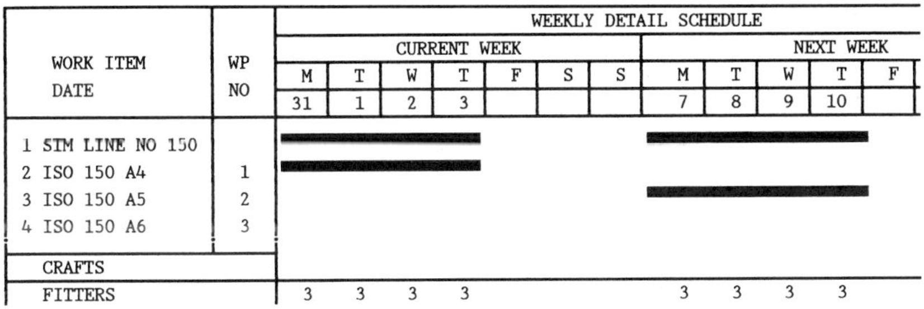

Fig. 4 - Contractor Weekly Detail Schedule

Work Statusing Techniques

In the context of this paper, work statusing techniques are ones used for determining the amount of work in place. Work in place is normally expressed in percent. This percentage reflects the proportion of work completed to the total work to be done. Once the work in place has been determined, earned value and productivity calculations can be made. One contractor uses the following techniques for calculating work in place:

Discrete:	Subjective:
Units Completed	Supervisor Opinion
Incremental Milestone	Cost Ratio
Level of Effort	
Start/Finish Percentages	

Units Completed - This technique is best used with easily quantifiable bulk items such as wire, concrete, steel, and pipe.

To calculate the percentage of work in place based on units completed, divide actual units measured by total forecast units.

For example,

$$\text{Percent Complete} = \frac{\text{Actual Units to Date}}{\text{Total Forecast Units}} \times 100 \quad . \; . \; . \; . \; . \; (1)$$

$$= \frac{4000 \text{ LF}}{10000 \text{ LF}} \times 100 = 40\%$$

This method is useful in that it is accurate, objective and easily audited. A disadvantage is that collection of data tends to be costly in terms of manpower. Another disadvantage is that the calculations need to be updated when the scope of work (total quantity) changes. Further, Thomas argues that the use of this method should generally be limited to fabrication shops which mass-produce bulk items and are better equipped to collect the required data [Thomas86].

Incremental Milestones - This technique is appropriate when the effort required to determine the percent complete at various stages of a task do not justify the effort to collect data. Frequently, the percentages allowed at each milestone are based on manhours required but in some cases the percentages are arbitrarily established. For example, an account for piping might consist of the following manhour estimates.

	Mhrs	Cum %
Received and Inspected	300	15
Setting Complete	400	35
Alignment Complete	300	50
Internals Installed	500	75
Testing Complete	300	90
Accepted by Owner	200	100
	2,000	

Fig. 5 is a depiction of the incremental milestone method.

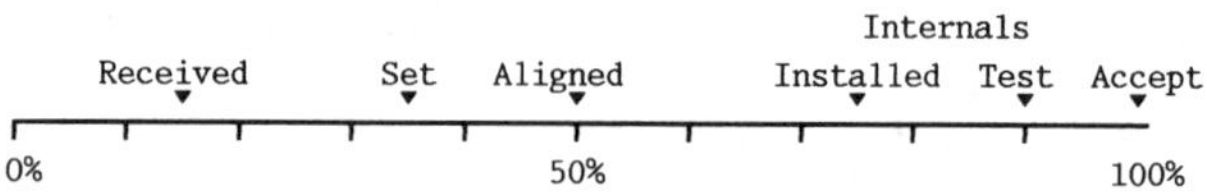

Fig. 5 - Percent Complete by Incremental Milestones

Level of Effort - This technique is used when the work consists of several subtasks and perhaps involves different trades. Fig. 6 is an example of a level of effort account. The advantage of this technique is its ability to handle diverse subtasks in one account. The disadvantage, as in the units completed method, is that calculations must be updated when quantities change.

M/H	Relative Weight	Subtask	U/M	Total Quantity	Quantity To Date	Eq Tons to Date
21	0.02	Run Foundation Bolts	EA	200	200	10.4
21	0.02	Shim	%	100	100	10.4
52	0.05	Shakeout	%	100	100	26.0
62	0.06	Columns	EA	84	74	27.5
104	0.10	Beams	EA	859	0	
114	0.11	Cross Braces	EA	837	0	
208	0.20	Girts & Sag Rods	BAY	38	0	
94	0.09	Plumb & Align	%	100	5	2.3
312	0.30	Connections	EA	2977	74	3.9
52	0.05	Punchlist	%	100	0	
1040	1.00	Total Control	TON	520		80.5

Fig. 6 - Level of Effort Account

Note: Eq Tons to Date = (Qty to Date x Rel Wt x 520tns)/Total Qty

Start/Finish Percentage - As discussed in the incremental milestone method, some tasks do not lend themselves to a rigorous analysis of intermediate progress. An example given was equipment alignment where the start and finish can easily be determined but measuring intermediate progress is difficult because of the uncertain nature of the work. Other examples include testing, painting, flushing and cleaning, and rigging.

In the Start/Finish technique, one might take credit for the mobilization effort by assigning some arbitrary percentage to the start of the work task. For example 20% could be taken when an item of equipment is delivered which signals the start of work. The advantage of this method is clearly its simplicity. Its disadvantage is the absence of intermediate checkpoints.

Supervisor Opinion - This subjective technique is recommended when more exhaustive methods are not justified. The technique requires a knowledgeable supervisor estimate the percent complete based on his observation and experience. Examples given included dewatering, landscaping, planning, and architectural trim.

This technique is simple and inexpensive. Its shortcoming is that it cannot exclude the bias of the supervisor and should therefore be used only for non-critical portions of a project.

Cost Ratio - This technique is suitable for tasks which are active over a long period of time and no attempt will be made to establish intermediate milestones. Examples are project management, quality assurance, and contract administration. The cost ratio percentage is given by:

$$\text{Percent Complete} = \frac{\text{Actual Cost to Date}}{\text{Estimated cost at Completion}} \quad . \; . \; . \; . \; . \; (2)$$

This technique is easy to apply. Its limitation is that it does not give adequate early warning of potential overruns.

Earned Value - All the contractors discussed in this paper used some form of Earned Value system. Earned Value is a method to measure progress in a project with dissimilar units of measure by converting these units of measure into either manhours or dollars. Thus a project "earns" manhours or dollars as progress in the various cost accounts is measured by one of the percent complete methods described above.

Generally, Earned Value is given by the expression:

$$\text{Earned Value} = \text{Percent Complete} \times \text{Budget For That Account} \quad . \; . \; . \; (3)$$

In Eq. 3, percent complete would be determined be one of the methods described above. It is noted, parenthetically, that some sources use the term Budgeted Cost of Work for the expression represented by Eq. 3 [DOE82].

For the account described in Fig. 6 above, the percent complete would be,

$$\frac{80.5}{520.0} \text{ x } 100 = 15.5\%$$

The budget amount is 1040 manhours and the Earned Value would be,

$$0.155 \text{ x } 1040 = 161\text{Mhrs}$$

Earned Value may also used to calculate the percent complete of the project as a whole:

$$\text{Percent Complete} = \frac{\text{Total Earned Manhours or Dollars}}{\text{Total Budgeted Manhours or Dollars}} \quad . \; . \; (4)$$

One contractor uses the above form of Earned Value to measure performance using a Cost Performance Index:

$$\text{CPI} = \frac{\text{Earned Manhours or Dollars}}{\text{Actual Manhours or Dollars}} \quad . \; . \; . \; . \; . \; (5)$$

In this case performance greater the 1.0 is favorable since the account is earning at a rate faster than actual expenditures. Another contractor used the following expression for Earned Value:

$$\text{Earned Value} = \text{Budget Unit Rate x Actual Quantity} \quad . \; . \; . \; . \; (6)$$

$$\text{Where Budget Unit Rate} = \frac{\text{Estimated Manhours or Dollars}}{\text{Actual Manhours or Dollars}} \quad . \; . \; (6a)$$

Although Eqns. 3 and 6 might appear to be different, they are essentially the same because if Eqns. 6 and 6a are combined, the result is, using manhours for simplicity,

$$\text{Earned Value} = \frac{\text{Estimated Manhours}}{\text{Estimated Quantity}} \text{ x Actual Quantity} \quad . \; . \; . \; (7)$$

Rearranging the terms slightly,

$$\text{Earned Value} = \frac{\text{Actual Quantity}}{\text{Estimated Quantity}} \text{ x Estimated Manhours} \quad . \; . \; (8)$$

which is the same as Eq. 3 if the units completed method is used to calculate percent complete. Where this second contractor differs, however, from the first is in the performance calculation.

For example the second contractor defines performance as,

$$\text{Performance Factor} = \frac{\text{Actual Manhours or Dollars}}{\text{Earned Manhours or Dollars}} \quad . \; . \; . \; . \; . \; . \; . \quad (9)$$

Note that Eq. 9 is the reciprocal of the CPI given by Eq. 3 and therefore performance less than 1.0 is favorable for the second contractor. Although certainly not a major issue, it points out a lack of standardization of terms among the owners and contractors in our study. Further, there seemed to be no groundswell of enthusiasm for standardization because the company procedures were well established, familiar to the users, and served company needs.

Trend Analysis

Productivity Profiles - Several types of productivity profiles were evident. The first and shown below as Fig. 7, represents how productivity varies over the life of a project. These curves might be used for the total project or for a readily measurable category of work such as cubic yards of concrete. For more detail, see [Clark85]. In the latter case, productivity would be measured:

$$\text{Productivity} = \frac{\text{Budget Unit Rate}}{\text{Actual Unit Rate}}$$

And, if productivity at 30% completion equals 1.06, then the actual percent of budgeted manhours used is 30.0/1.06 = 28.3%.

% Budgeted Manhours Used										
Theoretical	9.7	18.2	26.3	35.4	44.6	54.5	64.2	74.8	86.5	100
Actual	10.4	18.7	28.3							

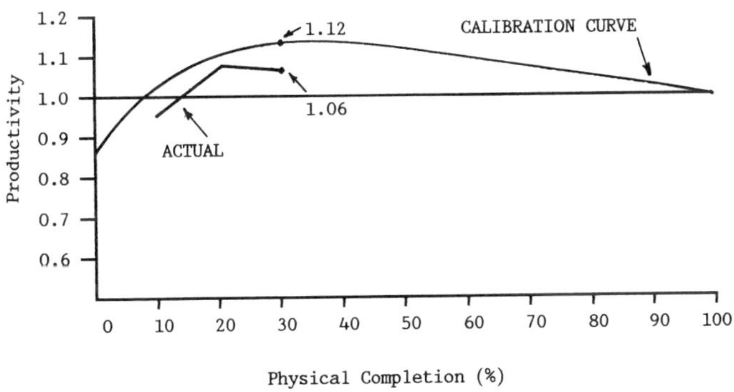

Fig. 7 - Project Productivity Profile

Shown below in Fig. 8 is another example of trend analysis using productivity charts. In this case productivity is measured as manhours per unit. For knowledgeable readers, 30 Mh/Ton may seem high, but these are the numbers actually used by one of the owner companies and perhaps include items not normally found in this category of work.

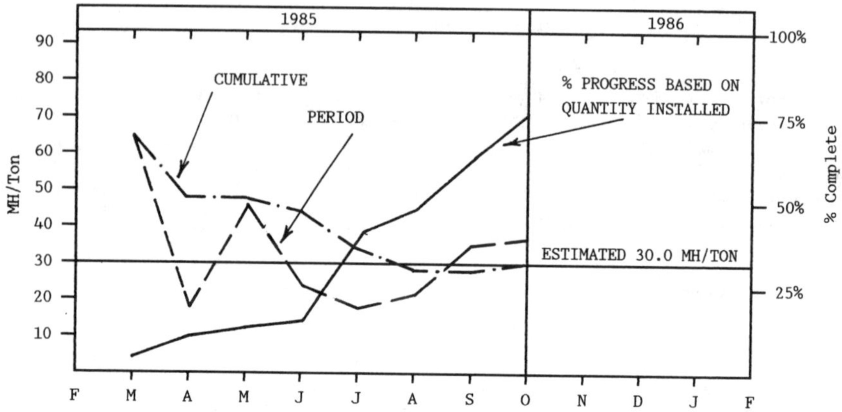

Fig. 8 - Structural Steel Productivity

Variance Analysis - All contractors and owners in this paper used variance analysis. Variance analysis is simply comparing current performance with expected performance. Typical items compared are

Budget Cost versus Actual Cost
Budget Unit Cost versus Actual Unit Cost
Budget Manhours versus Actual Manhours
Budget Percent Complete versus Actual Percent Complete

Variance analysis should highlight those items requiring special attention. A Department of Energy Guide notes that one should examine contributing factors to the variance such as poor initial estimates, technical difficulties, unexpected labor or material cost, and differing labor efficiencies [DOE80]. Variance analysis is a useful technique, but like the other techniques presented here is not meant to be used alone.

Progress Curves - These curves, commonly called "S Curves" are used extensively by both owners and contractors covered in this paper. Typical S Curves are shown in Fig. 9. Normally, these curves are plotted by "early start" and "late start." But, in Fig. 9, the curves represent a target schedule and a most probable schedule. These curves are used by an owner company. Although these curves are not unique, their use is interesting.

This owner has an extensive data base correlating project size (direct field labor manhours) and duration. Their analysis of com pleted projects revealed the contractor's initial estimate of construction duration was met only 30% of the time.

Therefore, early in the project, the owner expects the contractor to establish a schedule with a somewhat shorter duration than the owner's most probable. The owner calls this the "target" schedule. The contractor is then expected to meet the target provided expenditures are not excessive.

The owner also recognizes the impact on their operating division if commitments are made based on a target schedule with only a 30% chance of success. Therefore, they advise the operating division to make their business decisions on the basis of the most probable completion date.

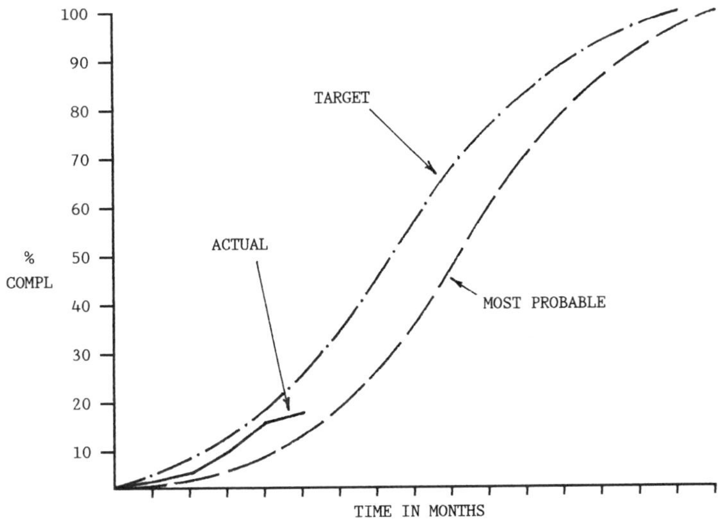

Fig. 9 - Typical S Curves

The owner then exercises schedule control by plotting actual progress against both the target and most probable. Toward the end of the job (80 - 85%) complete, past performance and expected future performance is reviewed and a consensus reached on a single control schedule. This schedule then becomes the basis of control to achieve the new predicted completion date.

Other progress control charts were noted in our study, but they were all essentially based on the concept of tracking progress against a cumulative "S" curve similar to those in Fig. 9.

Forecasting

Numerous forecasting techniques were noted in our study of owners and contractors. Three representative techniques are presented here.

The first technique is attributed to an owner company. This technique is used with the productivity profile shown on Fig. 7. At any given period, the actual productivity is compared with estimated productivity and the difference is used to forecast final productivity. The following example illustrates the method. On Fig. 7, at 30% complete, the actual productivity is 1.06 and the estimated productivity is 1.12.

Assumptions implicit in this method noted by Clark are that no intervention strategies will be used to change the outcome and that the percentage difference relationship will hold to the end of the job [Clark85].

Estimated Productivity		= 1.12
Actual Productivity		= 1.06
Difference		= 0.06
Percent Difference =	0.06/1.12	= 5.4%
Final Productivity =	1.00 - .054	= 0.946
Predicted Manhours =	50,000/0.946	= 52,854

Another forecasting technique noted is presented below. Partial data for cable tray are shown in Table 1 below.

		ESTIMATE		TOTAL TO DATE			
COST CODE		QTY	MH	QTY	MH	EST$/MH	MPF
DESC	UM	PROJ	COST	%	COST	ACT$/MH	CPF
6220		4000	3080	3990	3112	9.50	0.99
CABLE TRAY	LF	4200	29280	95	29875	9.60	0.98

Table 1 – Cable Tray Status Report

Projected manhours are calculated as follows. The Manhour Performance Factor (MPF) is defined as:

$$\frac{\text{Estimated Manhours}}{\text{Estimated Quantity}} \div \frac{\text{Total-To-Date Manhours}}{\text{Total-To-Date Quantity}} \quad . \; . \; . \; (10)$$

$$= \frac{3080}{4000} \times \frac{3990}{3112} = 0.99$$

Although Eq. 10 may not be readily familiar, it is in the form used by this particular company and gives the same information as Eq. 5.

And, Estimated Manhours to Complete

= Manhour Performance Factor x Remaining Quantity

= 0.99 x (4200 - 3990) = 208 MH

Projected Manhours

= Estimated Manhours to Complete + Manhours to Date

= 208 + 3112 = 3320 MH

The projected cost is calculated as:

Estimated Cost to Complete = Estimated Manhours to Complete x Actual $/MH

= 208(9.60) = $1997

Projected Cost = Estimated Amount to Complete + Cost to Date

= $1997 + $29875 = $31872

A third forecasting technique similar to that used by several companies is described in [Harrison83] and is adapted below. Cumulative data for the period are

BUDGET COST TO DATE	ACTUAL COST TO DATE	TOTAL BUDGET COST
7650	8370	21000

A Cost Performance Index is defined as

$$\frac{\text{Budget Cost}}{\text{Actual Cost}} = \frac{7650}{8370} = 0.914$$

Budget Cost for Remaining Work = 21000 - 7650 = $13350

$$\text{Estimate to Complete} = \frac{13350}{0.914} = \$14606$$

$$\text{Forecast Amount} = 8370 + 14606 = \$22976$$

A general comment regarding the three techniques described above is they are linear projections and, except for the first technique, do not take into account the variation of productivity over the life of a project. Additionally, none of these projections is attempted until the project is at least 25% complete.

Summary and Conclusions

A common theme observed among the companies in this paper was they all had excellent data bases. These data bases had been developed through years of experience on similar type projects. Meticulous maintenance of the data appeared to be a key factor in the companies success with cost and schedule control. The data base also permitted sophisticated planning methods in the early stages of a project.

Owner companies tended to control only key labor and material items. They felt their analysis of the data had shown the items that needed to be managed for effective control. Contractor companies tended toward more total control of cost and schedule.

Among the owner companies there appeared to be no consensus on the merits of integrating cost and schedule. On the other hand, contractor companies seemed to feel that integration was a worthwhile effort.

Both contractors and owners used work breakdown structures to plan and control projects. Contractors, as might be expected, made extensive use of work packages for field control.

Trend analysis was usually based on productivity or performance calculations. Forecasting, in turn, was based on productivity calculations. Additionally, most of the forecasting techniques observed were linear projections. The one technique which attempted to remove some linearity was the example shown in Fig. 7. Although this removed some of the linearity from the projection, the limitations noted in the methods above remain in this method as well.

Because so much is at stake on the conclusions drawn from forecasting, there seems to be a need for a more rigorous treatment of this topic. Perhaps simulation or risk analysis techniques could provide better results.

The increased use of job site personal computers to augment cost and schedule control was clearly evident. In some versions, the personal computers were used in a standalone mode and, in others, they were used in conjunction with a mainframe computer in the home office. The most prevalent use of these personal computers was in spreadsheet applications for cost control. We noted instances of internally developed software, but the preference was decidedly for off-the-shelf items.

An instance was noted in this paper of a lack of standardization of terms with regard to performance factors. Among the companies observed, this lack of standardization was widely evident. All the companies used their own systems and, although there was some similarity in methodology, the systems were essentially unique. For those readers interested in software development, finding a common denominator of terms and methods acceptable to a large number of companies would seem to be a relevant challenge.

Finally, although we observed many excellent and sophisticated applications of project controls we did not find any breakthroughs in the state-of-the-art nor did we expect to. To reiterate, our purpose in this research was to compile and report on the effective techniques currently in use.

Appendix — References

[Business82] "Modern Management Systems," Report A-6, The Business Roundtable, New York, November, 1982.

[Clark85] Clark, F. D., and Lorenzoni, A. B., "Applied Cost Engineering," 2nd ed., Marcel Decker, New York, 1985.

[DOE80] "Cost and Schedule Control Systems Criteria for Contract Performance Measurement," DOE/MA-0087, U. S. Department of Energy, Washington, May 1980.

[Harrison83] Harrison, F. L., "Advanced Project Management," Gower Publishing, Hants, England, 1983.

[Neil82] Neil, J. M., "Construction Cost Estimating for Project Control," Prentice Hall, Englewood Cliffs, 1982.

[Riggs86] Riggs, L. S., "Cost and Schedule Control Techniques in Industrial Construction," A Report to the Construction Industry Institute, University of Texas at Austin, December, 1986.

[Thomas86] Thomas, H. R., "An analysis of the Methods for Measuring Construction Productivity," A Report to the Construction Industry Institute, University of Texas at Austin, May, 1986.

Early Quantitative Scope Definition and Update on Fast-Track Projects

James T. O'Connor[1] A.M. ASCE

Conrad G. Vickroy[2]

Abstract

Poor scope definition and control have been identified as frequent contributing factors to cost overruns on industrial projects. The causes of inadequate scope definition are complex and deep-seated, and the effects are felt throughout the project duration. These causes and effects are explored in this paper, as are the closely related issues of the budget estimate process, the tracking of bulk material quantities, and the control of changes.

Investigations have led to the conclusion that product market pressure is the single most significant reason for underdeveloped scoping. Further, many project managers have grown complacent with this problem which is nurtured when factored cost estimates serve as the basis for control. The practices of delayed bulk material tracking and ineffective change control also fuel the scope control problem. As combative measures, project managers are advised to reduce the importance of factored estimates, to require designers to take-off quantities as part of their design activity, and to rigorously screen and account for changes. Many other recommendations are also presented.

Introduction

There has been considerable alarm in the business community over industrial project cost overruns in recent years. The Business Roundtable's Construction Industry Cost Effectiveness (CICE) Project Report A-6 (Business Roundtable, 1982) states that "Poor scope definition at the estimate (budget) stage and loss of control of project scope rank as the most frequent contributing factors to cost overruns." Another report (Smith, 1983) found that some construction industry officials consider lack of scope definition to be the most serious problem on construction projects.

While inadequate scope definition may be attributable to poor planning and management of projects, it must be assumed that there can be circumstances which preclude full scoping of projects or force managers to proceed at a level of scope definition less than they would like. Regardless of the reasons, when there is poor scope definition, final project costs can be expected to be higher because of the inevitable changes which disrupt project rhythm, cause rework, increase project time, and lower the productivity and morale of the work force.

---

1 Assistant Professor, Department of Civil Engineering, The University of Texas, Austin, TX 78712

2 Vice-President, Duane Whitlow & Co., Inc., Plano, TX 75075

This paper will explore the subject of scope control and attempt to answer the following questions:

Why does inadequate scope definition persist?
How do estimating practices sustain poor scope definition?
Why are bulk materials a particular problem?
How do changes relate to scope management?

Why Inadequate Scope Definition?

Industry Opinion: Lack of scope definition contributes to higher project costs and remains a common problem on industrial projects. Why are owners willing to proceed on the basis of inadequate project definition? Interviews held with owners and contractors provided explanations along these lines:

- Owner doesn't have engineering expertise to provide complete conceptual definition.
- If the initial order-of-magnitude (feasibility) estimate looks very favorable, there is no reason to spend additional funds in scope definition.
- As a matter of economy, owners want to limit the amount spent on feasibility and project authorization/budget studies.
- The shorter the time spent on preliminaries, the quicker the job can be completed and the lower the overhead.
- High interest rates make project duration critical.
- Market pressures make project duration critical.

Further investigation found that both owners and contractors consider market pressure to be the single most important reason for underdeveloped scope. The following explanations were offered:

- Demand for a product, while intense, may be short-lived. Many marketing analysts have noted a trend toward short product life cycles, particularly in technology-intensive industries.
- Competition in product development has increased. A representative from a company in a particularly competitive industry noted, "if we don't hit the market first, there may not be a market at all."
- In many cases, profits to be derived from plant operation seem to greatly outweigh changes in construction costs. An example cited was the Alaska pipeline where poor scope definition plus regulatory and environmental problems caused the project cost to rise from an estimated $900 million to more than $7 billion. In spite of this, the project is considered a success and profits have overshadowed all cost overruns.

In support of the market pressure argument, most industry representatives contend that further definition of scope would add to project duration (up to 25%).

Of the other possible reasons for inadequate scope definition previously noted, based on follow-up interviews, only two were given much endorsement and neither were ranked in significance with market pressure. The first is the lack of, or less than full use of, engineering capability within the owner organization to conceptually define a project. While it was commonplace in past years for large owners to maintain an in-house engineering and construction management capability, there is now a definite trend away from this as a matter of cost-effectiveness. Among those owners who still maintain engineering and/or construction divisions, many operate these divisions as reimbursable, independent organizations and are inclined to economize in their use because they are reimbursable.

The second reason relates to the euphoria that can accompany early cost estimates and revenue projections that are extremely favorable. In their haste to realize these revenues as quickly as possible, owners compromise good planning and scope definition.

Making the Best of the Situation: Ideally, owners should insist on full scope definition before proceeding with a project. But, for one or more reasons, owners can be expected to proceed on some projects with underdeveloped definition. One approach which may permit earlier and fuller development of definition is to have a stronger marketing-construction interface to permit earlier identification of potential market needs and their consequences in terms of construction. Of course, this is subject to the ready and continuous availability of engineering expertise. If such expertise is not maintained in house, an owner could consider establishment of an *evergreen* relationship with an engineering firm with experience in the owner's type of work. Over time this firm would become very familiar with the owner's philosophy and processes and be able to respond quickly to owner requests for conceptual engineering and budget estimates.

If time does not permit full scope definition, the risks of cost/schedule overruns should be part of the feasibility analysis of the project. If the decision is still to proceed, efforts must then be focused on scope control. Better procedures for developing realistic budget estimates, better procedures for quantifying bulk materials requirements, and better systems for controlling changes are the areas which can contribute most to scope control. The remaining sections of this paper concentrate on these areas.

Scope Definition and Control vs. Budget Estimating

The Problem: During the course of a project, a number of cost estimates are normally prepared. Although terminology is not standardized, the first estimate can be called *order-of-magnitude* and is usually part of a feasibility study. It utilizes cost/capacity curves or other scale of operations approaches related to output. It can vary by as much +/- 50%. The next level of estimate is the *factored* estimate. This estimate requires the preparation of flow diagrams so that major items of process equipment can be identified. It is then assumed that cost, quantity and dimensional relationships exist between this equipment and other project components. Costs are factored using these relationships. This estimate will have variation in the range of +/- 30%. The order-of-magnitude and factored estimates are *conceptual* estimates. The next level of estimate, the *control* estimate, is prepared once the piping and instrumentation diagrams (P&IDs), one-line drawings, and equipment plot plan/elevations are available. Quantities can be approximated from these items; and where definition is lacking, allowances are added. Accuracy should be +/- 15%. The highest level of estimate is *detailed* or *definitive* and is made once detailed drawings are available. The accuracy of this estimate should be within +/- 10/%.

Good scope definition is not considered possible until the P&IDs are available. Thus, budget estimates should be of the *control* estimate category. Unfortunately, many budget estimates are prepared before P&IDs are available and are of the factored estimate category. Consequently, the budget estimate is subject to the high variability previously noted. Much of the variability problem is associated with *offsites* (access, utilities, and other area-sensitive costs). These are more a function of geographic location, plot plan and arrangement than of equipment. As stated in one study (Merrow, 1981), "Most of the variation found in cost estimation error can be explained by . . .the degree of definition of the project's site and related characteristics."

Compounding the potential for conceptual estimating error on the low side can be the pressure exerted upon estimators to support inaccurate estimates by project sponsors who seek the go-ahead for their project. Project sponsores are understandably motivated to produce low cost figures to insure final authorization. If confronted with a high estimate on a project, they may challenge the estimators to substantiate it. Unfortunately, in conceptual estimating, judgment must replace straight forward material takeoffs so an estimator may find it difficult to quantitatively justify his/her estimate.

Another factor contributing to poor conceptual estimates is the inability of estimators to predict the direction of design development. They are not directly involved in the conceptualization and development of the design yet must prejudge those who are. They must work within the limits of whatever database is available and their personal experience.

The inexperience of many conceptual estimators may also contribute to the problem. Conceptual estimating is an art, not a science. It also tends to be an unpopular career goal. As one estimator has explained, "Many good people don't want to do it; it's a dead-end job."

Estimates as Control Vehicles: In a control system, the estimate establishes the baseline for control. Factored estimates are of little value for use in a control system despite the fact that they provide an estimate of total project costs and may provide rough quantity estimates. Their variability of +/- 30% is really applicable only to the *total* cost. Individual components within the estimate may have greater variations and, for this reason, their estimates make a poor basis for any type of control. An estimate based on P&IDs is the first estimate with potential as a control estimate. Table 1 is a summary of problems encountered with factored estimates. Recommendations are also presented.

Scope Definition and Control vs. Bulk Materials

The Problem: The installation of bulk quantities accounts for about 75% of field labor costs on an industrial project so such quantities have a significant cost impact. Until all drawings from a project are completed and bulk quantities taken off, material and labor costs associated with bulk quantities remain uncertain. In addition, since construction progress is often measured by ratio of bulks installed to total bulks, poor bulk definition inhibits accurate progress measurement. The effects of bulk quantity overruns surface late in the project -- in the construction phase -- yet the decisions that determine bulk quantities are made in the engineering phase.

Table 1. Factored Estimate Problems and Recommendations

Problem (1)	Recommendation (2)
Offsites impossible to factor accurately	Do not attempt to factor offsites. Make the effort required to perform a rough take-off which will require extensive utility and site investigation.
Operations pressure on estimators	Formalize company procedures, giving guidelines on ranges or factors to use. Expand factors to cover broader range of equipment sizes. Document conceptual estimator's judgement and/or reasoning when it is needed. Document differences between current and reference projects; have estimators show predicted impacts. Use some means to elevate status of conceptual estimators and increase rewards for accurate estimates.
Poor prediction of design development by conceptual estimators	Employ risk analysis techniques by identifying not only the most likely course of events, but also identifying and assessing the impacts of other possibilities. Develop a procedure for conceptual designers to sign-off on the information provided to the estimator. This will force estimator-designer interaction and make the designers less likely to stray from early intentions.
Importance of factored estimate	Minimize the use of the factored estimate for budgeting. The budget estimate should be of the control estimate category. Be willing to spend the additional funds needed to develop scope to the point that a more detailed estimate can be developed, and a control budget established.
Difficulties in use of factored estimate for control	Break the estimate into distinct areas; use same work breakdown logic that will be used for control later in project. Continually refine as design becomes further defined. Employ methods to reduce variability in the factored estimating process such as successive estimating.

Quantity definition and control at various stages is discussed below:

1. **The Factored Estimate Stage.** The factored estimate might be broken down into broad accounts, such as linear feet of pipe or tons of steel. These quantities are typically based on historical ratios of quantity per piece of installed equipment. The quantities generated in factored estimates should not be used for control purposes because the categories are so broad and significant variations are commonplace.

2. **The Control Estimate Stage.** The control estimate is generated after the P&IDs, electrical one-lines, and the equipment plot plan/elevations have been developed. By this time, significant definition of bulk quantities has occurred. The control estimate is based on material take-offs to the extent possible, plus allowances for items yet to be designed and contingencies. While this estimate provides some basis for control there are difficulties. With broadly defined allowance accounts there is a tendency to not recognize early overruns. Also allowance accounts of a broad nature tend to be identified as convenient *management reserve funds* or *contingency funds*. The solution for this problem is to break out allowances in more detail (allocated against individual accounts) and manage them account by account.

3. **The Definitive Estimate Stage.** The definitive estimate is prepared once detailed drawings are available. Even though engineering is considered complete, the field may still be required to route small bore piping lines, electrical conduit, or other bulk materials. To assure that the cost of this work is part of the estimate, an allowance for it should be included under a field control account.

Communication Difficulties: Communication difficulties within the engineering/construction process may complicate bulk quantity estimation and control. For example, differences may exist in the way different companies or different groups within the same company control piping. While material control may take off and count each piece of pipe and fitting, the estimating group may use linear feet to be installed of each type of pipe as the basis for its estimates while the project controllers may use a weighting approach that tracks all components but converts them to equivalent feet of pipe. The problem is compounded when the project is engineered by one party, while construction management and construction is handled by others. The solution is to standardize the quantity measurement methods used for estimating, statusing, and reporting. Once this occurs, the historical database will prove to be much more useful in preparing estimates for future work.

Design Engineering Involvement: Participative management is a well-established technique for increasing management effectiveness. The idea is that individuals involved in the development and operation of a program will be more committed to its support. In the case of bulk materials control there is potential for using this technique to improve that control by involving those who generate the quantities (design professionals) in the control process. It has been found that there is a correlation between effective control of bulk materials and the direct involvement of design personnel in their estimation.

Too often, design engineers are not required to pay attention to the bulk material quantities that their designs require. It is suggested that they itemize material quantities on the drawings they produce, both conceptual and detailed. A standardized indexing or parameter system can be used for bulk quantity estimation during early design stages; direct take-off is possible from detailed designs. If the designers cannot perform the take-off because of time pressure, material control personnel or cost engineers can be placed in the design office to do this estimation and coordinate the results with the design engineers. Automated material take-off programs are now appearing in conjunction with computer-aided design systems and should be considered. These will greatly speed and simplify the process. Expert systems also have promise in this area.

Another benefit accrues from making a designer responsible for the quantities he/she generates. By requiring immediate take-off, the length of the *cost blackout period* between design and the time its effect on cost is noted is essentially eliminated. If an overrun were detected immediately, redesign could be considered. With traditional systems, by the time a large increase in quantities is discovered, the drawing may have been approved for construction, materials ordered, and installation started. Redesign at this time would halt progress and is often not cost effective.

A summary of bulk material control problems and control tools is provided in Table 2. Details are contained in Ref. 3 (O'Connor, 1985).

Scope Definition and Control vs. Changes

The Problem: As noted earlier, loss of scope control during engineering is believed to rank second in impact on cost overruns. This is often referred to as *creeping scope.* The relationship between poor scope definition and the propensity for creeping scope and changes within scope is quite clear -- a poorly defined scope does not provide a clear baseline against which changes can be evaluated as being either changes within or outside of scope.

Changes within scope are often referred to as *project development changes* and include anything required to provide a facility as originally described. Anything else is a *scope change.* Unfortunately, there is no clear boundary between the two and the owner, as the buyer of the facility, must be deferred to in classifying a given change.

With a well-defined project, a good project organization, and an effective control structure, changes are readily classified and controlled. Project development changes can be value-engineered and otherwise evaluated and tracked against appropriate control accounts. Change requests which affect the scope of the project are obvious and are particularly scrutinized and challenged.

On a poorly-defined project, the impact of a design change request is difficult to classify, quantify, and justify. Poor project definition precludes development of a clear baseline upon which estimates and judgments can be made. Problems are compounded when control systems are not well enough developed to reflect cost impacts in short enough turnaround.

The cost effect of a change depends on the point in time it is introduced. If made very early in the conceptual engineering phase, there may be only the direct costs of the change to account for. At any later time, there can be a ripple effect on other engineering work and there may be construction rework and delays. The consequential effect tends to increase exponentially with time into the project.

Table 2. Control of Bulk Material Quantities

Problem (1)	Recommendation (2)
No quantities to control against in estimates.	Develop systems which will take factored quantities from estimates. Consider the use of an expert system developed from historical experience or a commercial estimating system designed for that purpose.
	Break project into work breakdown structure areas and develop quantity estimates according to that breakdown.
Monitoring during design development is difficult and may not be timely	Increase designer awareness of the importance of bulk quantities by (1) using designed quantities as a measure of progressing engineering, (2) requiring designers to take-off quantities as part of their design, or (3) using bonus incentives for quantity savings.
	Use first detailed drawings off the board as a sampling tool to check accuracy of early estimates.
	Continually expand and refine work breakdown structure and use as basis for estimates and control.
	Make systems for estimating and control of work compatible.
	Take advantage of automated take-off features available with computer-aided design.
Difficulty in monitoring and controlling field-engineered of quantities	Study engineering allowances and make them very specific to avoid consumption a field control allowance account before finishing engineering.
	Give responsibility for field-routed materials to the construction phase for better control.

Change Sources: Changes originate from several sources, the first being the *design engineers* themselves. These professionals must continually balance process and structural considerations with regulatory, maintainability, servicability, and human factors considerations; and, being human, will occasionally overlook something. Then, there is the problem of multiple disciplines working in parallel, each hoping their design work will be compatible with that of the others. In this atmosphere, there is considerable potential for design interferences, congested designs, and boundary problems, all of which often lead to changes.

A *safety committee* will review industrial facility designs for compliance with safety, environmental, and health requirements. Changes can result from these reviews.

Another source of changes is the *operating and maintenance staff*. These are the people that must live with the complete facility and they should have a voice in its design. However, several guidelines should apply to the use of their advice. First, they must be brought on board early in the conceptual design process. The later their ideas are introduced, the more costly will be any changes -- it is particularly dangerous to wait for their input until the construction phase. Second, their ideas should not be accepted carte blanche. Is the change just for someone's convenience or because that is the way it was always done in the past? Is the idea cost effective; is there an equally effective way that is more cost effective? Will these personnel actually be the ones assigned to the facility? If not, one can expect those actually assigned to have ideas of their own.

Constructability reviewers are another source of change. Ideally, constructability is taken into consideration during the conceptual engineering phase so that this knowledge can be reflected in the first designs. Still, the need for some constructability changes may not be evident until later and these must be formally evaluated as potential changes.

Facility sponsors are a major source of changes. This is particularly true on cost-reimbursable projects. For some of these projects, scope is considered to be not defined until the project is complete. As stated by one industry representative with respect to these projects, "there is a continuous design process; design is never *frozen*." Of course, many owners choose the cost-reimbursable route because it affords them the freedom to make changes freely if they are willing to absorb the extra costs. For them, operating, maintenance, safety or other requirements totally overshadow construction costs, and of course there is no real incentive for design contractors to control owner-requested changes on cost-reimbursable projects. Problems are compounded when owners have inadequate systems for filtering change requests from their own personnel.

Change Control: Whether changes are project development or change of scope, there is a need for a positive control system within an engineering organization. Suggested elements in a change control program are:

1. **Owner Commitment/Team Effort.** As with all programs, if change control is not supported from the top, there is little potential for its success. Without this commitment an owner should expect expensive changes and cost overruns. A formal change review committee may be an appropriate representative of this owner commitment. Such a committee could serve as a clearing house for changes before they are given to the engineering organization. A *team* atmosphere among owner, engineering and contractor participants is also promoted with this committee.

2. **Freezing Design.** The freezing of design is obviously the most effective way to control scope growth changes. Most owners and engineering

organizations agree that the time to do this is after the P&IDs have been completed. People will program themselves to comply with established deadlines, but will procrastinate if the door is left open. Owners and project managers must be absolute in their enforcement of this policy.

3. **Formal Change Justification.** With this approach, every change must be justified and reviewed by supervisors and affected parties. The initiator of the change must estimate the impact of the change both as to cost and schedule as well as other areas of the project that may be affected by the change. While this approach requires considerable effort and paper work, it does force considered thought of every change. Another major advantage is that it forces design engineers to become cost- and schedule-conscious.

A variation on this approach is the *selected change justification* in which there are two categories of changes. The first includes all changes with no identifiable alternatives. The costs of these changes are monitored closely, but high level approvals are not required. The second category includes changes with possible alternatives. All must be evaluated, one recommended for approval, and approval received from a designated executive. Affected managers must be alerted for both types of changes and only authorized personnel should be allowed to introduce changes.

4. **Operating Facility Liaison.** Some owners have found it beneficial to utilize the designated plant manager as liaison to the engineering organization during engineering. Since this individual is interested in both construction and operating costs, he/she can assist in decision making where there must be a trade-off between construction costs and operating improvements. This liaison could also serve as a clearinghouse through which all owner change requests would flow.

5. **Optional Changes.** Some changes are optional but appear to be a good idea because they provide savings in material or other direct costs. But, with every change there can be considerable consequential effects. One owner has adopted a rule-of-thumb to account for this that says that any change requested after P&IDs have been approved should provide a cost savings equal to 10 times the direct cost required to implement it. However, if the idea that the cost of change can vary exponentially with time of introduction is accepted, that ratio should probably rise to 25:1 or higher in later stages of detailed design.

Summary and Conclusions

The problems caused by poor scope definition and scope growth are felt by nearly every activity and in every phase of a project. This paper has examined the three areas most heavily affected -- budget estimate accuracy, bulk quantity control, and change management.

The only way to reduce the problems associated with poor scope definition is to adopt an organized, efficient philosophy for planning and controlling of projects. Table 3 outlines recommendations for improved scope control during the various project phases. The recommendations, if followed, should allow owners to operate under market pressure while also bringing positive cost and schedule control to their projects.

Table 3. Recommendations for Improved Scope Control

Pre-Conceptual Engineering

1. Identify the market for product in terms of how much is needed, when it will be needed, and how much revenue it can generate. Maintain capital cost data related to potentially marketable products.
2. Identify needs to execute project in terms of available company personnel versus needed personnel or services from outside the company.
3. Recognize the inaccuracy of the cost estimate in the initial feasibility analysis. If the analysis is sufficiently favorable, authorize development of flow sheets and needs for equipment.
4. Before proceeding to conceptual engineering, a letter defining project objectives should be drafted. The project scope should be limited to activities which support the listed project objectives.
5. As soon as possible, an operations representative should be designated as liaison between engineering and operating organizations.

Conceptual Engineering

1. Analyze duration, and cost impacts if conceptual engineering phase is lengthened to include reasonably complete piping and instrumentation diagrams.
2. Offsite information should be gathered at this phase to allow for a more accurate estimate.
3. Development of a work breakdown structure which will make it possible to control design development and changes later in the project should begin here.
4. Owner should be formulating change control policies, design freezes, and the responsibilities of the engineering organization in these areas.
5. Qualifications of the estimate produced at the end of this phase are needed. Definition of what is covered by contingency, and other items whose meanings vary should be addressed.
6. Estimators should be protected from operating pressures on their estimates.

Detailed Engineering

1. Work breakdown structure should be in place.
2. Formalized change control procedures should be instituted.
3. Engineers must be aware of the bulk quantities they are designing into the project.

Construction and Start-up

1. Ensure that design quantities are the same as those which are used for labor productivity measurement.
2. Ensure accuracy of as-built drawings. Incorrect as-builts on revamp projects are a source of unpredictable changes.
3. Make a comparison between early estimates and final costs. Document the assumptions in early estimates which were incorrect, and assess impacts. Update historical cost database with final cost values.

Acknowledgements

This paper is based upon work supported by the Construction Industry Institute and administered by Task Force 83-6, Cost/Schedule Controls. Particular gratitude is extended to Task Force Chairman J. M. Neal and member Jack Browder.

References

1. The Business Roundtable, *Modern Management Systems*, Report A-6, November 1982.

2. Merrow, E. J., et. al., *Understanding Cost Growth on Pioneer Process Plants*, The Rand Corporation, 1981.

3. O'Connor, James T. and Vickroy, Conrad G., *Control of Construction Project Scope*, University of Texas, a report to the Construction Industry Institute, December 1985.

4. Smith, Mark A., and Tucker, Richard L.; *An Assessment of the Potential Problems Occurring in the Engineering Phase of An Industrial Project*, University of Texas, a report to Texaco, Inc., 1983.

DESIGN ENGINEERING AND PROJECT CONTROL

James E. Diekmann[1]
K. Bryan Thrush[2]
Thomas A. Wilson[3]*

INTRODUCTION

This paper describes the results of a Construction Industry Institute (CII) funded study which investigated procedures to accomplish project control on design projects. One primary goal of the study was to define HOW major design firms applied cost and schedule control techniques to their design projects. The second goal of the study was to defined WHICH techniques appeared to be the most successful in controlling cost and schedule.

The state of the practice in design engineering project control was assessed through interviews with designers and owners. Six design firms and eight owner organizations were interviewed between August and December of 1984. All fourteen companies are involved in heavy industrial and process plant construction. The interviewees included project controls managers, scheduling managers, cost managers, project engineers, and designers. By interviewing both the designers and those that commission the design both views of project controls were examined and compared.

An interview guide was developed to provide a framework for each interview and to insure that similar data be collected from each interviewee. The guide was in the form of a questionnaire that covered specific project control topics as well as general information concerning the designer/owner and how the firms conduct business. The guide was divided into six main topics:

1) - General
2) - Organization
3) - Work Breakdown Structure/Chart of Accounts
4) - Design Activities and Progress Measurement
5) - Project Control Systems
6) - Historical Performance

The first group of questions were designed to determine the type of work performed and who was responsible for controlling that work. Opinions were requested from the interviewees concerning what factors influence the success or failure of a project and the attitudes within

---

* [1] Associate Professor, Civil, Environmental and Architectural Engineering, University of Colorado - Boulder

[2] Field Engineers, Stolte, Inc.

[3] Estimating Engineer, Aluminum Company of America

the organization concerning project controls. The second area of questions centered upon the corporate and project organizations. The purpose is to determine the status of project controls personnel and to determine the project controls organization. Attempts were made to define small, average, and large projects and the corresponding control staff.

Next, the interviewee was asked to describe the Work Breakdown Structure or Chart of Accounts utilized to separate projects into manageable pieces. Questions were asked concerning the number of levels of breakdown and the corresponding work packages.

Once all the necessary background information was obtained discussion focused on the elements of the project control system. The fourth topic of discussion concerned defining each type of project activity and how progress of each is measured.

The last but most extensive group of questions focused upon the firm's project control system. Scheduling philosophy and methodology, cost monitoring, and reporting procedures was explored. Each element of the project control system was examined in detail.

The owner interviews focused upon project control requirements owners place upon designers and what is done by the owner to monitor the design contractor.

All of these data were then compiled, classified and studied in order to discriminate between those control procedures which are effective and those that are not. In this paper we will present the major findings from this study but first we attempt to convey the primary lesson which we learned from the study.

PRIMARY FINDING

This study is the result of the efforts of project managers, design managers and project control personnel with several major design firms and large owner's organizations. The study required many hours of interviews, too many airplane rides and untold hours of sifting and analyzing data. The study also benefited from the moral and financial support of CII and their Cost/Schedule Task Group. Given all of the time and effort contained in this project, it is difficult to reduce these efforts to one principal finding. However, adopting contributions from two widely divergent fields may help to express the primary lesson on Design Project Control that emerged from this effort.

The first contribution is a quote from the linguist Alfred Korzybski...

"The map is not the territory"

In our efforts to determine schedules, calculate earned values and explain variances, we build a model (a map) of the project. The model, if it is skillfully built, helps project personnel to navigate the route to the end of the project. Yet no matter how skillfully the model is constructed, the model is not the project. ("The map is not the territory"). Missing from the model is the vital component, people. Successful project control systems are designed such that they are used by and useful to people. In this light, whether one uses arrow diagrams or precedence diagrams is not very important. Whether one has five levels or six levels in a WBS is immaterial. System must be tools for people.

The second contribution is from the physicist Fritjof Capra who wrote a book entitled "The Tao of Physics", in which he tries to reconcile eastern mysticism and western science. Although there is much more to the book, one idea is repeated throughout the book. This idea is that "reductionism" eventually is no longer a useful paradigm for scientific inquiry. The reduction paradigm seeks to explain phenomena by reducing it to its constituent parts. Thus, a molecule is reduced to atoms, and atoms to protons and electrons. Yet, there is a point where further reduction does not lead to further understanding. In the realm of sub-atomic physics it appears that one must take a holistic view to understand the world of quarks and anti-matter. In project control, it is common to use reduction to help explain project performance. Design activities and performance are described in terms of numbers of drawings and specifications. Drawings are described in terms of various milestones. But as with sub-atomic physics, at some point project personnel must abandon the reductionist view and adopt a holistic view of a project. At some point, it is essential that the various measures and sub-measures of project performance be combined with the _experience_ and _insight_ of the senior project management to establish the condition and circumstance of the project. Clearly, it is not enough to use a project manager's intuition and feelings to measure the status of a project. Just as clearly, it is not enough to rely solely on mathematical calculations and computer graphics. In some manner, successful project control systems must allow their human users to include their experience and insight in the output of the control system.

If there is one fundamental finding from this study it is this:

PEOPLE CONTROL PROJECT COSTS AND SCHEDULES, SYSTEMS DO NOT.

OTHER FINDINGS AND CONCLUSIONS

Findings Related to Organization

Finding 1 - Organizations which emphasize project responsibility are more effective.

Organizational structure may be classified as one of three types - a functional organization, a matrix organization, or a project (task force) organization.

The purpose of an organizational structure is to define an orderly communication network. The channels must be arranged in a timely and efficient manner and be compatible with every project. Because of the large number and size of projects undertaken by designers in the construction industry, matrix organizations are common. If a matrix organization is to be used, which variation is best - i.e., a weak or strong matrix? A weak matrix gives the functional department greater authority than the project manager. This arrangement can create conflicts for the project participants. A strong matrix organization gives the strongest authority to project management - those who are responsible for producing a technically sound product that meets cost and schedule restrictions. A strong matrix also facilitates timely and efficient communication channels by placing each team member in the same locale. A strong matrix also builds a stronger "team approach" needed to tackle a project. Strong matrix organizations keep project objectives more clearly focused. On very large design projects task force organizations, in which members sever all associations with functional departments, have been used to good advantage.

Finding 2 - There is an increasing emphasis being placed upon educating project personnel in project controls and on decentralizing project control groups.

In order for a project control system to be effective it must be accepted by all who use it. If the individuals using the system have an understanding of how and why a project control system works and also are given a say in the input to that system, they will be more inclined to accept and use the system.

Finding 3 - Project control systems are not designed to support the working level.

All designer's project control systems are basically designed to report information to management. The output from any system is only as good as the input. Since project personnel are providing the input to the project control system, the system must be designed with the ability to produce daily information for project personnel. If discipline engineers use the project control reports for their daily activities, they will be certain to provide good input information. Systems must be created or current systems modified to make project control information more meaningful to discipline managers.

Finding 4 - Project Managers control the support given to project control.

The emphasis upon project controls is dependent upon the project manager's viewpoint. Therefore, project managers should be well-educated and trained in the importance and use of a project control system. Such is the case with most, but not all, project managers.

Finding 5 - Designers are expending a similar amount on project control activities.

It is a difficult task to determine what amount of control is enough. Based upon the designers respondents, it would appear that approximately 8% of the design budget is an appropriate amount. However, every project is unique with its own combination of problems, 8% expenditures may be sufficient on one project and inadequate on another. A cost/benefit appraisal should be performed by each designer to determine the proper amount of time and effort to be expended on project control activities.

Findings Related to Planning

Finding 6 - 'Up-front' planning is a major key to a successful project.

All designers agree that project pre-planning is a pre-requisite for a successful project. But much planning is enough? Often the amount of planning spent by the designer is dependent upon the amount of information contained in the bid package released by the client. The amount also depends upon the size and nature of the project as well as the project team chemistry. Generally, designers did not have a prescribed system or procedure to accomplish early planning. The general consensus of those interviewed was that the amount of pre-planning was insufficient on most projects.

Finding 7 - Scheduling philosophy is similar but methodology varies.

The majority of designers interviewed develop schedules in a hierarchical manner. The hierarchy begins with a summary project schedule. This schedule outlines major project milestones only. Generally, it contains between 10 to 50 activities and is in bar chart form. Differences arise in the breakdown of the summary project schedule into the lower levels. The number of activities, the area designations, and the use of a bar chart or CPM logic network all very considerably among the designers.

A hierarchical approach to scheduling is recommended because it affords a logical sequence to development. It also allows a smaller number of activities per schedule without sacrificing detail, making schedule control a much easier task. Scheduling in a hierarchical manner is conducive to controlling schedule by work packages. There is a trend among firms to use some kind of work packaging scheme to control schedule and those who do seem the most satisfied with their scheduling procedures and results.

Also, engineers should be given more responsibility in developing schedules. More realistic schedules are produced and engineering is more committed to keeping that schedule because it is their schedule.

Finding 8 - There is a decline in the value placed upon CPM scheduling.

The consensus among the interview participants is that the CPM is not an effective tool in scheduling the details of engineering work. Most firms used a CPM schedule at the second level which is used for interface with procurement and construction. Below the second level however, the trend is away from CPM scheduling and toward scheduling using list, logs and spread sheets. There is increased usage of bar charts supported by detailed lists and data bases of work to be accomplished.

Findings Related to Progress and Performance Measurement

Finding 9 - Progress Measurement methods are similar but embody different philosophies.

The greater the amount of work that is progressed formally, the more effectively the project may be controlled. It has been demonstrated by two designers that the procurement effort can be formally progressed. It has also been demonstrated by one firm that other non-document activities such as cost/scheduling and modeling may be progressed. Effort should be expended to develop progress measurement system for procurement activities and as many other activities as possible.

Finding 10 - Vendor data and vendor drawings are often a stumbling block to engineering production.

Due to the project impact that vendor information has on engineering production, some measure of control over that information should be established. The vendor print effort must be given appropriate high level attention otherwise design schedules are jeopardized by lack of design, procurement and vendor communication.

Findings Related to Control

Finding 11 - There is a trend toward controlling work by work packages or groups.

Controlling work by work packages is an effective method in project control. Controlling by work packages improves the information that is processed by the project control system. Work packaging produces units of work which are meaningful to both the design forces and the construction forces.

Finding 12 - Short interval scheduling is an effective tool to control schedule.

Short-interval scheduling is generally accepted by the industry as an effective tool to control schedule. It allows engineers to concentrate on the present by setting short-term goals and makes it much easier for engineers to plan their work.

Finding 13 - Project control systems must be designed to handle the small jobs as well as the large ones and also should be standardized within a company.

Both of the above-mentioned problems must be dealt with. First, firms should develop a standardized project control system. Only through the continued use of a standardized system will users learn to effectively use it to control work. It will also eliminate the charge to the project cost that occurs every time a new system is utilized.

Second, this standardized system must be able to handle the smaller jobs. It should be able to provide timely information for the small project's management. If it is not possible to use one system for the entire range of project sizes encountered by a firm, then as a minimum, the same control philosophies must be used in every system used by a firm.

Finding 14 - Small computers are the next technological step in project control.

The benefits of small computers are many. It was stated that project control systems are not geared enough to the working level. A small computer strategically placed in the hands of project personnel can solve this problem. They are easy to use and provide quick access to project information. Software in the way of scheduling packages, cost control packages and integrated cost/scheduling packages is available for small computers. With the trend toward keeping project information to a manageable level and the trend toward controlling work by work packages, projects are lending themselves to being capable of being controlled by the small computers.

Finally, small computers are a less costly alternative to project control. Small computers cost only a fraction of that of a large computer and cost very little to run. By placing a number of small computers in the hands of project controls groups and discipline managers timely and efficient control can be fostered at the working level at relatively low cost.

Finding 15 - Changes are one of the most troublesome aspects for a project control system.

Project control system and control are based upon planned budgets and schedules. When changes are made to the original plans, the planned schedules and budgets are no longer an accurate standard to measure progress by. Therefore, the planned budgets and schedules must be revised to reflect those changes. But often this task becomes very complex. Changes to one part of a project may effect many different departments' budgets and schedules.

By the time all of those affected groups revise their designs and drawings to accommodate that change it may end up costing more than the original expected savings.

Changes also place a strain upon the communication system. Changes must be channeled to the affected departments. Approval for changes must be debated and if approved all affected budgets and schedules must be revised. It may also force project personnel to change their plan of work.

Finding 16 - People, not systems, are the key to a successful project.

For a project control system to be effective, the people using the system must understand and accept it. They must understand the importance of it as well as understand how to use it. If it is understood that the project control system is a tool to help the user plan and control his work rather than a tool that only tells him (and management) he is doing something wrong, then it will be accepted and used.

A concentrated effort should be made to focus attention on the project personnel. They should be educated in the use of project control systems and given more responsibility in using the system. The project control system should not take too much "care and feeding" by the project discipline engineers but should remain in the background as much as possible. The project control function must be viewed as an integral part of the overall project effort rather than a separate chore or task to be accommodated.

Findings Related to the Owners Impact on Project Control.

Finding 17 - The perception of the role to be played by the owner in project control is similar but the application of that role varies dramatically.

While all owners perceive their role in project control to be one of giving direction and monitoring the progress of the designer, their application of that role varies dramatically from one of heavy involvement to one of only summary-level involvement.

Those owners who exercise close involvement seem to be the most satisfied with their project results. Designers also agreed that heavy client involvement (but not micro-management) was a key factor in successful projects. Heavy owner involvement can make the designers task of project control easier and should be a prime consideration in future endeavors.

It is the owner's interest to walk the fine line between heavy involvement and micro-management of the project. Owner involvement promotes timely decisions and improved communications. It engenders cooperative attitudes among the project team. Micro-management is the situation where the owner is not only making owner decisions but also the designer's decisions. Micro-management causes designers to feel that their perogatives and responsibility are being taken away.

Finding 18 - The amount of 'up-front' planning and information provided in the bid package differs dramatically from firm to firm.

The amount of planning performed by an owner before and during a project varies greatly. It is difficult to determine the cost associated with increased planning efforts versus the benefit in total project savings, but is generally accepted that decisions made very early in the project have the greatest impact on project costs. If the greatest cost impact occurs in the critical planning stage of the project it is plausible to assume that the sooner the scope is defined the lesser the chance for cost overruns.

Finding 19 - The trend is to partition projects into several smaller projects.

Separating projects into several smaller projects is advantageous to project control. From an owner's standpoint it makes the task of control easier by breaking the project into manageable pieces. From the designer's standpoint it make the task of control easier by reducing the amount of information to be processed and controlled by the project control system.

Finding 20 - Most owners concentrate on schedule.

Schedule control is one, if not the most important, elements of project control for an owner. This fact alone is enough to justify active owner involvement in the scheduling process. The owner is the one who knows what he wants and how he wants it done. By taking an active role in schedule development the owner can convey his expectations and wishes clearly and the designer can spend more time in production and control. The more guidelines that a designer has to start with the better the plan of work will be and in turn the easier the task of control. At the very least owners should provide the designer with a definitive project schedule. It is also recommended that the owner also get involved in the entire schedule development process.

Finding 21 - People are the key to project success.

This finding is so important that it is being restated. The best possible system imaginable will not control design project costs or schedules, people do. As one owner participant stated, "its the people that make a project go. No one system is better than another - its the people that make the difference."

CONCLUSIONS

It is difficult to present conclusions to a paper such as this for it is actually a string of conclusions. Yet, it is the writers' view that the construction industry must spend more energy on developing useful project control systems. Control systems which demand involvement of the key project personnel, and in return give to those personnel, information which will help them accomplish their responsibilities. Too often the project control literature is concerned with the advantages of a new software package.

Regardless the type of computer or type of software used, the system must be a "tool" which can be used by designers to control project cost and schedule. The system must be "friendly" enough so that designs can access the data they require. On the other hand, the system must yield consistent and accurate answers.

The genuine need is in determining how to best convince the lead engineers (and superintendents and other who are charged with project responsibilities) that cost and schedule control is a major part of their responsibility.

REFERENCES

1. "Modern Management Systems", Report A-6, November 1982, Construction Industry Cost Effectiveness Project, The Business Roundtable.

2. "Project Control in Design Engineering", January 1986 by James E. Diekmann and K. Bryan Thrush, University of Colorado, under contract to the Construction Industry Institute.

3. Project Control for Engineering, CII Publication 6-1, July 1986.

ACKNOWLEDGMENTS

The authors wish to acknowledge the generous support of the Construction Industry Institute in underwriting this study. The unselfish contributions of those employees of designer and owner organizations who participated in these interviews is gratefully acknowledged. Finally, the authors recognize the contributions made to this study by the Cost and Schedule Task Force of CII which is ably directed by Dr. Jim Neil.

DISCLAIMER

The opinions presented in this paper are solely those of the writers and in no way represent the stated position of the CII.

CURRENT NEEDS FOR COST CONTROL SYSTEMS

Paul M. Teicholz*

INTRODUCTION

Cost control systems are widely used within the construction industry. These systems normally produce weekly and monthly reports that summarize cost, budget, manhour and forecast information against a chart of accounts that represents a hierarchical work breakdown structure (WBS).

These systems are sometimes well integrated with accounting systems, but are rarely integrated with scheduling and material management systems. In addition, the errors in data contained in the cost system are rarely highlighted so that the bad data or omitted data can be corrected before the reports are used by management. This often results in wrong signals being given to management and a loss in confidence in the reporting system.

This paper reviews the desired integration between cost and schedule and material management systems to make them more useful. It also describes techniques for screening the data in a cost system for possible reporting problems, so that these can be corrected before a report is used.

INTEGRATION OF COST AND SCHEDULE SYSTEMS

The cost system utilizes a chart of accounts as the work breakdown structure (WBS) against which budgets are entered and all other data is collected. This WBS is developed when the project is estimated although modifications are normally made prior to the start of the project when more time exists to study project reporting requirements. The WBS reflects the combined (and sometimes competing) needs of project management, accounting, estimating history and owner information. For some types of work, e.g. buildings, there are standard codes defined by the American Institute of Architects (AIA) that can be used. For most other types of construction work, there are no standards and each company normally develops internal coding standards. A typical cost code would have a number of components to indicate:

---

*Corporate Director of Systems, Guy F. Atkinson Company, 10 West Orange Avenue, South San Francisco, California 94080.

Project	eg. Delray Plant Expansion
Area of Project	eg. Building #4
Major type of work	eg. Concrete forms
Detailed type of work	eg. Set and Strip 8" wall forms
Type of cost	eg. Direct Labor
Type of resource	eg. Journeyman carpenter

A fixed number of digits (or letters and digits) would be assigned to each portion of the code.

From the perspective of cost-schedule integration an important point is that a given cost code can often cover the same type of work done in a number of locations, eg. several areas within building #4. For cost control there is no need to separate each location since they would normally all be done by the same crew and foreman, and would all have the same unit budget (productivity). In fact, creating a separate cost account for each location could cause coding problems in the field since the foreman would have to allocate the time of his crew and himself to each location. This could be difficult if the crew worked on two or more locations in a given day. Inaccurate time allocations could easily occur.

The definition of network activities however must reflect the location of work, since the location is often the source of dependencies that impact the network. Thus, using the example above where a single account was defined for "set and strip 8" wall forms" for building #4, the network diagram could have separate activities for setting wall forms in each area of the building and stripping the same forms since these activities take place at different time and are planned individually.

For this reason, it is important that there be a flexible mapping of a cost account to the one or more activities to which it relates. There is also a possibility that more than one account can be related to the same activity. An example of this would be where an activity of "build 8" wall on east side of building #2" could relate to a number of cost accounts for each type of work needed to build this wall, eg. build forms, place rebar, set forms, place concrete, strip forms, cure concrete. Thus, a many-to-many relationship is needed between accounts and activities (see fig.1).

Figure 1 - Relationship between Cost Accounts and Network Activities

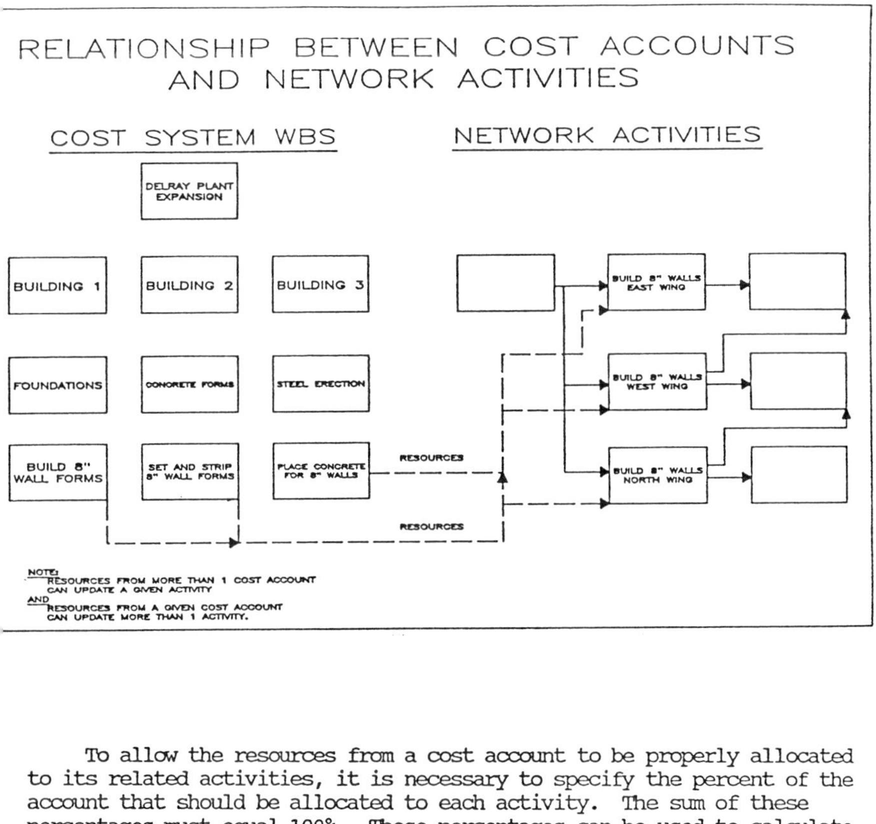

To allow the resources from a cost account to be properly allocated to its related activities, it is necessary to specify the percent of the account that should be allocated to each activity. The sum of these percentages must equal 100%. These percentages can be used to calculate the allocated values of budget as follows:

Let i = index on cost accounts
j = index on activities
r = index on resources ($, manhours, etc.)
BTOT(ir) = total budget amount for resource r at account i
BTOT(ijr) = total budget of resource r allocated from account i to activity j

F(ij) = fraction of account i allocated to activity j (assumed to be the same value for all resources) (between 0 and 1)

A(i) = set of activities related to account i

Then

$$BTOT(ijr) = F(ij) * BTOT(ir) \quad \text{(by definition)} \qquad (1)$$

Where

$$\sum_{j \in A(i)} F(ij) = 1.0 \text{ for a given activity } j$$

Allocation of to-date and remaining amounts is more complex, since it is necessary to take into account the status (completed, active, unstarted) of each activity related to a given cost account. Clearly, the to-date amounts of a cost account should only be allocated to completed and active activities and not to unstarted activities. Conversely, the remaining costs of an account should only be allocated to the remaining portions of active activities and unstarted activities and not to completed activities. This can be done as follows:

Let

P(j) = to-date time percent complete of activity j (between 0 and 1)
TD(ir) = to-date amount of resource r at account i
TD(ijr) = to-date amount of resource r allocated from account i to activity j
REM(ir) = remaining amount of resource r at account i
REM(ijr) = remaining amount of resource r allocated from account i to activity j
FTOT(ir) = total forecast final amount for resource r at account i
FTOT(ijr) = total forecast amount of resource r allocated from account i to activity j

Then

$$TD(ijr) = TD(ir) * k(ij) \qquad (2)$$
$$REM(ir) = FTOT(ir) - TD(ir) \qquad (3)$$
$$REM(ijr) = REM(ir) * m(ij) \qquad (4)$$
$$FTOT(ijr) = TD(ijr) + REM(ijr) \qquad (5)$$

Where

$$k(ij) = P(j) * BTOT(ijr) / \sum_{j \in A(i)} \big(P(j) * BTOT(ijr)\big) \qquad (6)$$
$$m(ij) = \big(1-P(j)\big) * BTOT(ijr) / \sum_{j \in A(i)} \big(1-P(j) * BTOT(ijr)\big) \qquad (7)$$

k(ij) is the fraction of the to-date amount to allocate to each activity. m(ij) is the fraction of the remaining amount to allocate to each activity. An example will illustrate these formulas (see fig.2.)

Account: Place concrete slab 10,000 sf
Budget = $0.30/sf x 10,000 sf = $3,000

To date qty = 4,000 sf = 40% complete
To date cost = $0.35/sf x 4,000 sf = $1,400
Remaining cost = $0.433/sf x 6,000 sf = $2,600[1]
Forecast Cost = $1,400 + $2,600 = $4,000

Related activities:

1. Place concrete slab - east wing 3,000 sf
   100% complete (based on time)
2. Place concrete slab - south wing 4,000 sf
   25% complete
3. Place concrete slab - north wing 3,000 sf
   0% complete

Total: 10,000 sf

	Total Account	Act 1	Act 2	Act 3
Work Quantity, sf	10,000	3,000	4,000	3,000
Allocation,%	100%	30%	40%	30%
Total budget,$	3,000	900	1,200	900
To-date Qty,SF	4,000	3,000	1,000	0
Percent complete	40%	100%	25%	0%
k(ij)	1.00	0.75	0.25	0.00
To-date cost,$	1,400	1,050	350	0
m(ij)	1.00	0.00	0.50	0.50
Remaining cost,$	2,600	0	1,300	1,300
Forecast cost, $	4,000	1,050	1,650	1,300

[1] based on a manual forecast of the remaining unit cost

Figure 2 - Relationship between a Cost Account and its Related Activities

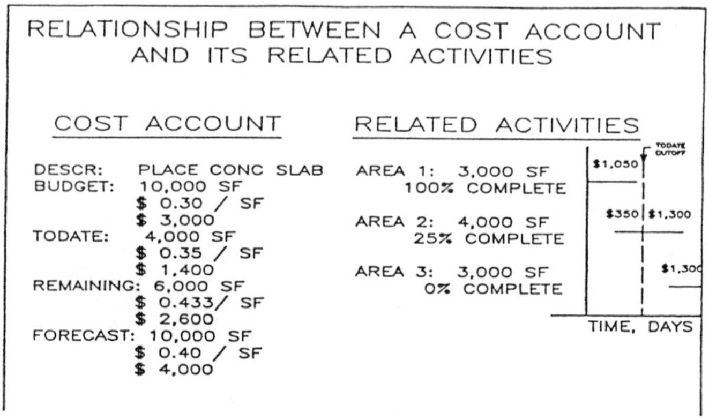

The above allocation of costs to activities is recognized to be approximate. These formulas assume a linear (uniform) distribution of resource usage over the duration of an activity. If this assumption is not valid, eg. a bell shaped curve is a better approximation, then formulas 6 and 7 must be modified by replacing the time percent complete of an activity, P(j), by the resource percent complete for a particular resource at that activity, P(jr). This percentage is based on an assumed resource distribution which is approximated by a specified histogram. For example, on a particular activity it is assumed that the costs will be 50% expended when the account is 30% time complete. This percentage (50%) would be a better value than the time percentage (30%) when calculating the fraction of to-date cost of an account to allocate to each of its linked activities.

Another caution should be mentioned regarding the use of these formulas. When the cost and schedule systems get out of synchronization, it is easily seen by reviewing the results of the resource allocations. The following problems could easily occur.

(1) A cost account with no remaining cost is tied to activities with remaining duration thus implying that no cost will be expended over the remainder of the its tied activities.

(2) A cost account with remaining cost is tied to activities with no remaining duration, i.e. all are completed. This is the opposite of case #1 above.

(3) After allocations for each account have been made, the resulting resource usages rates over the duration of activity may be quite misleading. For example, an activity with a budget manhour rate of 24 manhours per work day (3 man crew), could experience a to-date usage rate of 2 mh/day and a remaining rate of 10 mh/day (a 5 to 1 ratio).

All of the above situations would indicate that the linkages between cost accounts and activities were probably incorrect. The reporting system should highlight such cases so that they could be reviewed before the reports are used by management.

Despite the potential inaccuracies in the above approach, it is proposed as being preferable to an alternative that requires a one to one relationship between cost accounts and activities. The main advantage of a flexible relationship is that it allows both the WBS used by the cost system and the network activities used by the scheduling system to better serve their intended uses. The main disadvantage of a flexible relationship is that it requires careful monitoring of both systems to keep them properly synchronized.

INTEGRATION OF COST AND MATERIAL MANAGEMENT SYSTEMS

A material management system (MMS) is used for purchasing, expediting, tracking, inventory management, etc. The transactions associated with these functions include: requisitions, purchase orders, change orders, receipts, issues, invoices, inventory adjustments, etc. These transactions need to have their corresponding impact on the cost system. The necessary integration points are as follows:

(1) Purchase orders
-- update the proper cost accounts with the committed cost for each line item.

(2) Change orders
-- update the proper cost accounts with the change order amount for each line item.

(3) Invoices
-- update the proper cost account with the invoiced cost so that the remaining cost of each purchase order line item at this account can be calculated.

(4) Inventory Issues
-- update the issued to account with the proper cost (direct plus burden) and material quantity.
-- update the issued from inventory account with equal and opposite cost.

(5) Month end Accruals
-- At month end it is necessary to determine whether material has been received but not invoiced. It is necessary to accrue the cost of this material in the cost system to reflect the true financial situation of a project. If cash accounting is used (rare on large projects), this requirement does not apply. The MMS can automatically generate this cost accrual to the proper cost accounts because its input includes all receipts and invoices.

The above points of integration between a cost and MMS are often lacking in stand alone systems, thus almost assuring that these systems will not reflect a consistent picture of material status. It is most important that committed costs be automatically reflected in the cost reports where they serve to give an early warning of material and subcontract cost overruns, i.e. before these costs are invoiced. This gives project management the opportunity to avoid these overruns or plan on other cost reductions.

POTENTIAL ERROR ANALYSIS OF COST SYSTEM DATA

A cost system contains a large number of data elements, which are being frequently updated during a project. Such data as work quantities, labor manhours, committed costs, actual costs, forecasts, etc., are entered on a periodic or as required basis. Cost reports which present this data in a variety of formats are then used by management for cost monitoring and control. On a medium or large project, there is so much data that two problems must be addressed:

(1) How to screen the data to highlight potential errors which should be corrected before the reports are used by management, and

(2) How to highlight the most important areas for management attention in a clear and concise manner (management by exception).

Of these two problems, the first is often ignored while the second has received considerable attention (ratios, ranking by gain/loss/cost, etc.). This paper will present an approach called potential error analysis (PEA) that has been successfully used in a large construction company on a wide variety of projects.

The PEA for a given project consists of a listing of potential error conditions for each account. See fig. 3.

Figure 3 - Possible Error Analysis of Printed Accounts

6504 7302 COGENTRIX-ROXBORO MONTHLY TOTAL COST & EARN REPORT

POSSIBLE ERROR ANALYSIS OF PRINTED ACCOUNTS (MAX OF 25 ERROR

CONTINUED

CURNT PD COST WITHOUT CURNT QTY

ACCOUNT	CURNT COST	CUR WORK QTY
701-2300	25,792	
701-2000	18,158	
626-3100	4,088	
420-1000	3,491	
701-2200	2,290	
407-3000	1,980	

CURNT PD QTY WITHOUT CURNT COST

ACCOUNT	CURNT COST	CUR WORK QTY
613-5000		85
613-3400		2

CURNT PD UNIT COST > 25% DIFFERENT FROM TODATE

ACCOUNT	CUR UNIT COST	TDT UNIT COST	DIFFERENCE	% DIFF
626-1120	57.399	8.831	48.568	550.0
613-1500	21.473	4.638	16.835	363.0
626-1110	12.613	7.477	5.136	68.7
626-1100	64.080	40.901	23.179	56.7
614-4000	.358	.282	.076	27.0

MANUAL QUANTITY FORECAST REQD FOR NON LS ACCOUNT

ACCOUNT	BUDGET QTY	TODATE QTY	DIFFERENCE	% DIFF
613-2200	21,250	40,009	18,759-	46.9
626-2100	2,304	2,568	264-	10.3

AUTOMATIC COST FCST > 10% DIFF THAN PREV VALUE

ACCOUNT	CURNT FCST	PREV FCST	DIFFERENCE	% DIFF
626-1000	4,981	106,273	101,292-	95.3
701-2300	67,137	41,345	25,792	62.4
611-1500	84,624	59,058	25,565	43.3
613-1600	133,311	113,667	19,644	17.3
701-2000	102,797	84,638	18,158	21.5
613-2500	60,332	44,029	16,302	37.0
626-2100	52,317	36,640	15,676	42.8
613-3200	2,095	17,126	15,031-	87.8
613-2600	36,055	21,480	14,574	67.9
613-1500	42,181	30,019	12,162	40.5
613-3500	89,593	101,059	11,466-	11.3
515-1000	19,615	9,304	10,310	110.8
626-1100	24,540	15,695	8,845	56.4
626-1110	13,832	5,492	8,340	151.8
626-1120	12,274	4,679	7,594	162.3
613-3600	60,715	67,924	7,208-	1[illegible]
611-1700	7,152	2,002	5,149	2[illegible]
613-1200	36,832	31,883	4,948	
626-3100	13,987	9,898	4,088	
613-2200	10,365	6,395	3,970	
410-9000	5,541	2,191	3,350	
650-2100	3,979	1,305	2,673	
614-4000	15,560	12,974	2,58[illegible]	
614-3000	3,676	5,833	2,15[illegible]	

Examples of error conditions that are tested for include the following:

(1) Manual cost forecast > 10% different from linear cost forecast: (indicates that the manual cost forecast should be reviewed).

(2) Manual cost forecast is required for a lump sum account that is exceeding its budget: (indicates that to-date cost has already exceeded budget and that a manual cost forecast is needed to tell system what is a reasonable estimate of final cost).

(3) To date actual cost exceeds its manual cost forecast: (indicates that either manual forecast is in error or that actual costs have been improperly coded to this account).

(4) Current period cost without current quantity: (indicates that either a work quantity needs to be entered against this account or that current period costs have been miscoded to another account). This error is shown in Fig.3.

(5) Current period quantity without current cost: (indicates that either current period work quantity or cost have been miscoded). This error is shown in Fig.3.

(6) Current period unit cost > 25% different from to-date unit cost: (indicates that an error was made in measuring current period work quantity or that current costs were miscoded or that a significant change in productivity has occurred). This error is shown in Fig.3.

(7) Manual forecast quantity required for a non lump sum account whose to-date work quantity exceeds its budget quantity: (similar to #2 above, except applies to work quantity). This error is shown in Fig.3.

(8) Automatic forecast this period is > 10% different from forecast last period: (indicates that either current period cost or quantity is in error or that a significant change has occurred in productivity — similar to #6 above).

(9) Current period cost or quantity against an account completed prior to current period: (indicates either a coding error or that account should not be in completed status).

(10) Committed cost > actual cost for a completed account: (indicates that a change order which reduced a P.O. has not been entered or that invoice has been incorrectly coded, or that account should not be in completed status).

(11) Committed cost > manual forecast: (indicates that either a change order has not been entered or that manual forecast is in error).

Each cost account is tested for each possible error condition. To limit the number of potential errors generated, it is necessary to highlight only those that are significant, i.e. should be reviewed and possibly corrected. A given account can generate more than one potential error condition. The corrective action for a given error normally consists of reviewing the data in question and making corrections to one or more cost accounts, e.g. revising or entering forecasts, revising or entering progress work quantities, shifting costs that were coded to the wrong account, etc. These potential errors are normally difficult to find without a great deal of manual effort and it is a good task to automate. A significant number of potential errors in the database of a given project is a clear indication that insufficient attention is being paid to the cost system as a management tool and that corrective action is needed. In actual use, the PEA has proven to be a vital component of an effective cost system.

SUMMARY

This paper has presented three aspects of a cost system which are considered important and are often missing in practice. These include:

(1) Integration of cost and schedule systems using a flexible mapping between the cost WBS and network activities.

(2) Integration of cost and material management systems, and

(3) Use of a potential error analysis to screen and highlight reporting errors that should be reviewed _before_ cost reports are used by management.

These are issues that directly bear on the usefulness of a cost system as one of the management control systems in use on a project. There is an obvious need to integrate the cost, schedule and material management systems so that they reflect a consistent view of the project. In practice this is a difficult and often neglected task. It is hoped that this paper will help address some of the problems associated with this goal.

Assessing Changes and Their Real Impact

Weston T. Hester*
John A. Kuprenas**

Traditionally, changes on fixed-price contracts are negotiated on the basis of estimated direct and indirect costs. Unfortunately, the actual productivity and costs are seldom known, and the claimed costs are considered suspect at best. In this paper, a microcomputer-based system for gathering productivity and cost data related to changes and their impact is described, and its use on current projects is reviewed. The measured impact of common classes of changes on direct and indirect costs is summarized, and the manner in which this data was collected is described in detail. In conclusion, it is shown that costs are generally substantially greater than anticipated, but they can be effectively tracked and documented with a project-level cost control system.

Introduction

A great deal of attention has been given in the literature to the legal significance changes and to listing factors that may affect them. But, to effectively manage changes we must establish the associated increase or decrease in project time and money in a realistic and quantitative way. Yet, in practice, this is very difficult even after the work is completed. Typically, there are differences over what constitutes the original and new scope of work and what may be considered compensable. Owners frequently want to pay only the direct and immediate costs associated with a change, while contractors often demand this and additional compensation for reduced productivity, protracted delays on the project, and increased administrative costs.

The purpose of this paper is to briefly review a contemporary approach to documenting the time and cost impacts of changes, and then to present and illustrate an effective approach based upon the use of site-based microcomputers. Contractors and others may use the approach outlined here to effectively track the cost and schedule impacts associated with a change and establish a more realistic level of compensation for them.

*Associate Professor, Dept. of Civil Engineering, University of California, Berkeley, California 94720.

**Graduate student, University of California, Berkeley, California 94720.

In this paper we emphasize cost control and change management practices used by small to moderate-sized contractors (i.e., under $150 million annual volume) on small to moderate project sites (i.e., less than $40 million). The specific observations presented here were made on work performed by general and subcontractors.

A Standard Project Control System

The basic principles of project control systems are quite simple. The contractor divides the work into identifiable tasks and establishes a schedule for each task. The contractor also assigns cost codes to the work as a whole, and tracks the labor and materials required to complete any one task by cost code.

A representative construction task may be installation of a gypsum wallboard wall in a large commercial office building. The installation of the stud system would represent one cost code, fixing the gypsum board to the studs would have another cost code, and separate cost codes would be assigned to taping and texturing. But, all of these cost codes are combined to represent a single "task", i.e., the installation of a wall. Another construction task may be installation of fire sprinklers in a ceiling area, and cost codes be assigned to each activity associated with installation of sprinkler heads, straight runs of pipe, and support hangers. Labor hours expended by cost code for each crew are established by the respective foremen, these records are then summarized to form the project record for the period, and then a more complete, typically monthly, report is prepared for the project as a whole.

Crew foremen record the labor hours for each craftsman and estimate quantities of work installed on a daily or weekly basis for each cost code. A labor clerk then summarizes the time cards prepared by the foremen, establishes the total number of man-hours expended for the project by cost code, and estimates the quantity of work in place. This summarization of foremen records is normally a two step process, however. First, the clerk will prepare a summary of all of the data as reported, but must then rationalize any apparent discrepancies or errors in the data. Common errors include duplicate or missing time cards, incorrect assignment of work to cost codes, and gross errors in quantity estimates. Once the time cards and other subtotals are summarized at the project leve, all subsequent analysis is done at a remote administrative office.

The steps in a conventional, manual cost and schedule control system and one automated for the same project are summarized below in Figure 1.

In principle, this conventional, manual system works well. The contractor can divide the work into reasonable cost codes and accurately monitor progress for each of them. In practice, even when there are no changes the system tends to be excessively cumbersome.

The apparent workability of existing, manual control systems notwithstanding, there are at least three overriding flaws with them. First, workers often spend a disproportionate amount of time waiting for materials or directions with changes, but rarely have a cost code to charge this non-productive time, and thus do so innappropriately. Secondly, the summary reports are often returned to the site two to four weeks after the reporting period, and thus are too late for use as a short-term control tool. Finally, any given cost code will often be common to several tasks, and it is thus difficult to isolate any one task and its related costs.

Figure 1

Steps in the Collection of Cost Control Data:
Manual Versus Automated Data Analysis

Step 1: Prepare daily time sheet -- foremen summarize man-hours and quantities by cost code for each craftsman

Manual Approach:

Step 2: Weekly summary sheet -- foremen consolidate daily data into weekly report

Step 3: First consolidation by cost code -- labor clerk consolidates all foremen reports by cost code

Step 4: Final consolidation by cost code -- labor clerk corrects and consolidates all foremen reports by cost code

Step 5: Data summarized for monthly Project Report -- prepared at main office

Automated Approach:

Step 2: Daily summary prepared -- labor clerk inputs data

Step 3: Data input verified -- project manager verifies input

Any failings in the project control system are magnified if the project is subject to significant changes in the scope or nature of the work. Records for the cost and schedule impact of changes are normally kept in the same format and level of detail as the regular control system, with special notes referencing critical events and costs. As the project progresses, the net cost and schedule impact of each change is estimated and an agreement reached.

But if traditional project control systems suffer because of errors and summary reports returned well after the work has been performed, the ability to control changes is particularly impacted. When accurate records of the man-hours actually expended on a particular change are not available, or must await a periodic report delivered well after the task is completed, the contractor and owner cannot have a clear sense of the real costs and schedule impacts incurred. The lack of timely and accurate records is particularly insidious with changes because neither the contractor or owner will be able to fully appreciate the real cost increases associated with change until well after the work has been performed. And, of course, at that time it is too late to mitigaste the impacts of the change or to moderate its scope.

In conclusion, manual cost and schedule control systems may not be appropriate for projects involving poorly defined work or extensive changes. These projects particulalry need a mechanism for tracking the work costs and schedules on a daily or at least weekly basis, and these records must be immediately available to management personnel in the field and remote office. Manual systems, however, are not suited to daily updating of information and accurate summarization of the data reported.

Automated Cost Control

Use of a site-based microcomputer or terminal linked to a remote CPU permits the contractor to automate all of the intermediate steps in the summary of control data, as illustrated in Figure 1.

In addition to simple consolidation of the intermediate steps, automating the data analysis also provides at least three other significant benefits: (a) reduced errors in data input; (b) immediate feedback on costs to date; and (c) the opportunity to assess costs by tasks as well as cost code.

We have used the automated cost control procedure outlined above on four different projects. For a period of time up to several months, this automated procedure was used in parallel with the pre-existing manual system. Our particular concern was to evaluate the efficiency of the automated approach.

In projects we have studied, there is typically at least a 5 to 10 percent error rate with each intermediate step in the manipulation of the cost control data. As a guide, on projects ranging in size from 3 to 45 million dollars, when the data in the field was prepared mannually approximately 20 percent of the monthly summary report issued by the main office is incorrect. The net result is the reports are less credible, and particularly so at the project level, but time must still be spent to correct or justify the data used. With the automated process, approximately 5 percent of the report has a detectable error.

In addition to the benefits of accuracy, with summaries prepared on a daily basis, project management may immediately identify and assess long-term trends. In addition, by sorting the data input by foreman, management may quickly assess productivity for each task comprising several cost codes. Ultimately, then, this facilitates the evaluation of the impact of changes in the work.

In conclusion, objectively, automation of the data analysis reduces errors and the time required for report preparation, but it also has an important psychological component. When the data is collected at the project site but processed elsewhere, project management and particulalry foremen discredit the cost reports, tend to assign errors to others, and feel inhibited about using the data as a real management tool. When the data is input at the project site and analyzed with a microcomputer there, project personnel believe they have more tangible control over the data input and the resultant output. Thus, they are more likely to use the data to evaluate the productivity of alternate construction sequences and methods, and to assess the potential cost and schedule impact of changes.

Managing Changes With Automated Cost Control

Automating cost control has both a short- and long-term impact on the way changes are managed. In the short term, with immediate feedback on the work and productivity for a particular task, management can quantitatively assess progress in the work and the potential impact of changes in the working conditions, schedule, or other developments. In the long term, however, there is another benefit and one that may be overriding.

Since control records developed and maintained at the site are perceived as being more credible by the project personnel, we find there is greater attention to maintaining accurate records and to updating them on a timely basis. And, because data input is done daily, rather than weekly, the foremen are able to more precisely record conditions that affected the work and estimate the quantities in place. Specifically, the foremen are willing and able to more closely document changes and interferences in the work, and associated costs.

Automated System for Control of Changes

To monitor changes on the project site, we developed a simple, automated system based on the relational database, dBase III Plus, and performed work on Compaq, IBM PC and IBM AT machines with at least 512 kilobytes of RAM. A hard disk was available on some of the machines, but for convenience when working with different machines all data was kept on flexible disks.

All data was input on a daily basis from records maintained by the foremen. Figure 2 contains a summary of the information provided by the foremen and the information already input into a master file for the project. Note the project is divided up into a series of tasks comprising a number of common cost codes, and as the work is performed each foreman is assigned a task. Thus, costs may be summarized by cost code for the project or for a singular task and foreman. As the daily records from the foremen are input, other information needed for payroll and cost purposes is retrieved from the master file.

Figure 2

Contents of Input Forms and Working Files

Data provided by foremen:

- Cost code
  - Quantity of materials
  - Man-hours
  - Work area number
- Task performed
- Foremen's and craftsmen's names
- Code(s) for special conditions

Project Master File:

- Cost code
  - Description
  - Estimated quantity
  - Estimated man-hours
  - Estimated cost
  - Foremen
  - Assigned work areas
    - Assigned tasks
    - Payroll rates and data

Note that the only truly new information being recorded by the foremen are codes for special conditions. The Master File for the job keeps track of the foreman by work area and cost codes which the have been assigned that foreman, and thus allows minimal input of data by the foreman.

With respect to known or potential changes, the foremen have specific codes to record direct intervention by the owner to delay or change the scope or sequence of work, the presence of unanticipated conditions, and the presence of other personnel in the area. A list of approximately 8 such codes is used for all work.

As the data is input, a check is continuously performed to confirm that the foremen are using allowable cost and other codes and that recorded costs are within expectations.

Once the raw data is input, separate files are generated by the system, including one summarizing all costs and quantities by cost code and one summarizing all costs, quantities and other data by the type of change or other special condition.

For a work force of 40 to 50 craftsmen and foremen, we have found input and all automated file updating takes approximately 2 hours per week of a clerk-typist's time. In a side-by-side comparison of the time required to input data manually and in an automated manner, automated input was approximately 5 times as fast (as a trial, the two approaches were used simulataneously for about two weeks).

Several forms of reports are available upon request, including ones summarizing costs and quantities by cost code, changed versus unchanged work, and tasks or areas of work. These reports are a direct summary of the individual data files.

In summary, automating the processing of labor time records and quantities can significantly reduce the time required for the work and increase the accuracy of reporting. But, in addition, with little or no extra effort this system can also be modified to permit isolating the work of a particular crew or operation of a particular task. The application of this is illustrated in the next section.

Once the daily time reports have been collected, input and summarized in the system, several forms of analysis may be performed. For simple cost control purposes, the most common analysis is to examine actual versus budgeted costs for specific areas of work and tasks. It is, after all, the tasks being performed by a particular foreman, or work that is being done in an area, that is the focus of management attention.

To assess the cost impact of changes and their impact on productivity, other analyses are performed. This assessment may be done from printed reports, but is frequently done more clearly and quickly with Lotus 1-2-3 or other spreadsheet packages that provide graphical summaries.

The Impact of Changes on Productivity: An example

A classic problem in the management of changes is to assess their impact on productivity. Changes may involve unforeseen interruptions or other delays of the contractor's operations, forcing several contractors to work in the same area, or unforeseen events that make it more difficult for the contractor to have access to his assigned work area.

For one project, the work was divided into six basic tasks, and each could be done independently. Each task represented the installation of a particular detail in a mechanical piping system. Tasks that required special skill and care were grouped as <u>extra detailed</u>, and other work was classed as <u>detailed</u>. Together, both groups were referred to as <u>all work</u>. All project data was gathered and analyzed using the automated system described previously.

Many of the tasks needed to be performed high (30 feet) off the ground and in places where the work area was severely limited. The delivery and handling of materials were not constrained, but it was a somewhat precarious position for the individual craftsman. The average relative productivity for detailed and extra detailed tasks in the elevated location and on the ground level are summarized below in Table 1.

Relative productivity is defined as the productivity for a particular task in a particular area versus the productivity for the same task over the project as a whole. In Table 1, it is shown that craftsmen working in an elevated area are nearly twenty percent more productive, and cosistently so, relative to those working in a clear area on the ground.

Table 1

Relative (%) Productivity Versus Type of Work Area

Productivity for project as whole = 100

	Clear Area		Elevated Area	
	Detailed	Extra Detailed	Detailed	Extra Detailed
Sample Mean	99	99	117	119
Sample Std. Deviation	92	89	58	64
Number of Observations	155	96	10	6

Based upon the data presented in Table 1, project management attempted to assess why work in an elevated area was more productive. After discussing this data with the foremen and analyzing additional information, it became clear that two factors were responsible. First, prior to working in remote areas, the foremen made a special effort to assemble all tools and materials prior to starting the work, and the work itself was carefully planned. The time for this planning and preparation is explictly included in the time for each task.

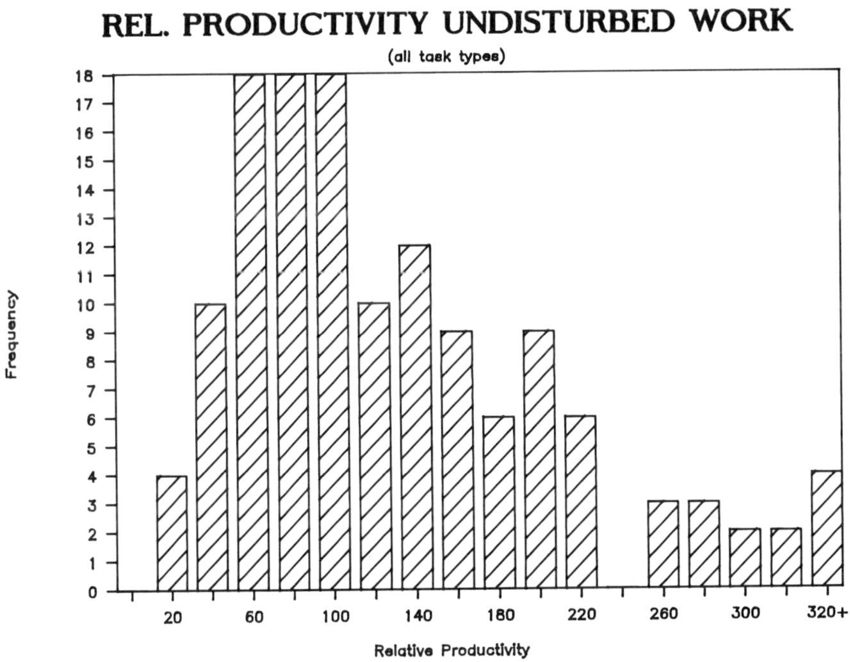

Figure 3

Frequency Distribution of Relative Productivity
for Undisturbed Work

Sample Mean = 123 Std. Deviation = 93 Number of points = 134

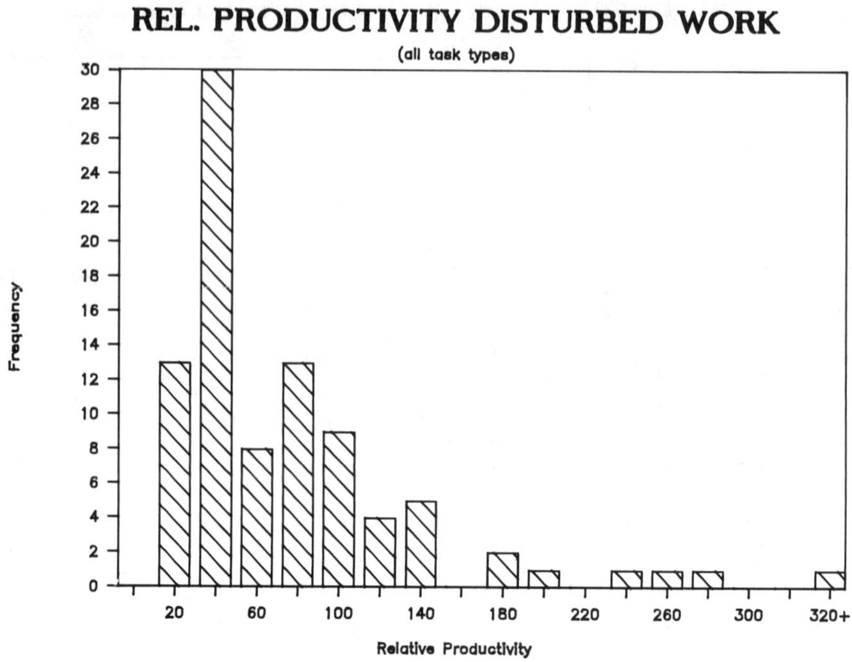

Figure 4

Frequency Distribution of Relative Productivity for Disturbed Work

Sample Mean = 65 Std. Deviation = 63 Number of points = 89

Secondly, the foremen reported that if a craftsman were working in a remote area he would not be interrupted to perform other work, and directed to return later. In our observations, the fact an interruption occurred was recorded, but the delay in returning to the work was not included in the time required to perform the task. In effect, the data in Table 1 suggests that interrupting the craftsman's work results in lower productivity, the length of the interruption notwithstanding.

Another concern on this project was the impact of owner-imposed interruptions in the work and congested work areas on the relative productivity. Work performed with more than a specified number of owner-imposed delays or in certain congested conditions was classed as disturbed work. The frequency distribution for different levels of relative productivity under undisturbed and disturbed work conditions is summarized in Figures 3 and 4, respectively. Note that work not disturbed is likely to be much more productive than the average level, and disturbed work is likely to be much less productive.

In summary, labor costs and material quantities reported by the foremen were tracked with an automated cost control system able to record and report on changes in the work. The control system was then used to identify productivity for specific tasks by nature of the work, levels of congestion in the work area, source of the change, and other indicators.

Conclusion

Until quite recently, small and medium-sized contractors developed and maintained project cost and quantity records manually on-site, with final payroll and accounting records computerized at a remote office. With the expanded use of microcomputers and the availability of flexible data base software, however, contractors are making increased use of computerized control systems on the project site. By properly setting up these control systems, contractors can also effectively track and assess the impact of changes. Furthermore, this can be done with little or no additional effort.

In the future, increased attention should be given to using data collected on productivity associated with different tasks and working conditions to more effectively schedule the work and minimize the disruptive influence of changes.

CONCEPTUAL COST ESTIMATING CONTROL SYSTEMS

Hans C. Björnsson*

ABSTRACT

This paper discusses approaches to conceptual cost estimating and their usefulness as cost control tools during the design and development decision-making process. An Approach that "simulates" the final design and applies relevant cost data to this design is found to be the most advantageous approach. Such a method must be developed on database concepts.

Introduction

Estimating or predicting what something will cost and then controlling the cost of doing that something within the limits of the estimate is an essential part of any project. Every project has a life-cycle made up of four phases:

- evaluation and planning
- conceptual engineering
- detailed engineering
- construction

During the first phase the main emphasis is placed on the screening of different alternatives with little emphasis on establishing an estimate of the project details. During the second phase, the conceptual engineering phase, the basic design is determined leaving enough flexibility for alternative solutions for different details. A project estimate therefore requires a number of assumptions and predictions. The "conceptual" estimate requires design knowledge as well as production knowledge in the application of historical data. It forms the basis of the first total-cost picture of the project. During the detailed engineering phase, the definitive control estimate is prepared. Project details become known and "accurate" costs can be predicted for the project. During the construction phase controlling the costs using the detailed estimate as a "standard" or budget is in the main focus. In this paper only conceptual cost estimating and its control function will be discussed.

*PhD, Director, Institute for Management of Innovation and Technology (IMIT), S-412 96 Göteborg, Sweden.

Definitions

An estimate is a guess or a prediction of how a project will be executed. The accuracy of the prediction depends on the quality of the estimation, the estimating methods used as well as the level of detail outlined in the design documents. The estimate will serve as the standard against which costs will be controlled. A conceptual estimate is an estimateperformed during the early design plans of a project. The design of projects have become a complex process, often complicated by schemes like fast-track construction in which design is only partially completed when construction starts. Such schemes obviously lead to considerable uncertainty, making the process of estimating and controlling a project highly risky.

Cost control is a management process wherein budget requirements are identified and compared with estimated project scopes in the context of costs, labor, materials, equipment and other resources. Project cost control has three dimensions: cost, time and quality (Kharbanda et al., 1980). These factors can all be expressed in monetary terms and are also interrelated and interdependent. Cost control refers to all methods of controlling these costs within the limits of a predetermined time, throughout the design and construction stages.

"Conceptual cost estimating control system" in this context is a system used in the conceptual project phase to establish the conceptual estimate and to control the design against this estimate.

The Importance of Cost Control

It is vital to operate an effective cost control procedure during the design stage of a project to keep the total cost of the job within budget. Effective cost control procedures enable the designer and the decision maker to be kept fully informed of the cost implicatins of all design decisions.

The main aims of cost control are (Seely, 1982):

(a) To give the owner good value for money
(b) To achieve a balanced and logical distribution of the available funds between the various parts of the building
(c) To keep total expenditure within the amount agreed by the client, based on an approximate or conceptual estimate of costs prepared in the early stages of the design process.

There is a need for strict cost discipline throughout all stages of design and execution to ensure that the intial estimate, the bid price and the final account sum are all closely related. Any influence at all on cost, let alone the controlling of costs, diminishes rapidly as the project progresses. The diagram, figure 1, illustrates this point graphically.

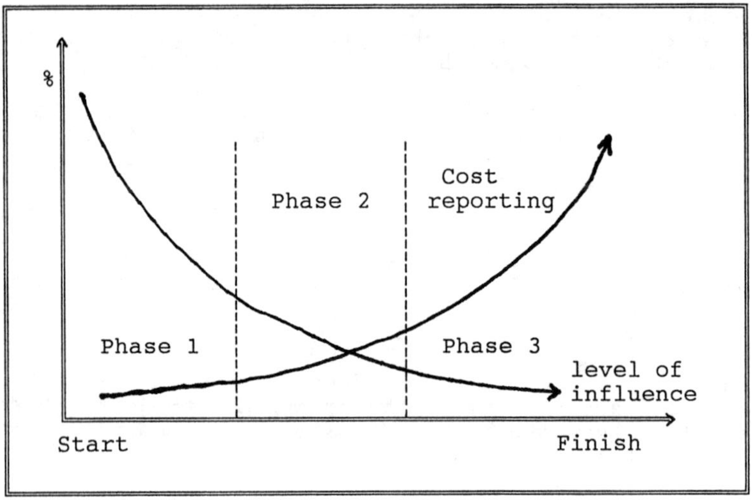

Figure 1 Cost control potential

As a general principle every project incurs little cost in early stages and much higher costs towards its end. The large expenses towards the end of the project follow automatically from decisions made early in the project. It is therefore important to be able to estimate the costs in early stages.

Conceptual Estimating: Establishing the Control Basis

Cost control has been defined above as all methods of controlling costs within the limits of a certain budget. The budget is established from an estimate accomplished from rough design ideas, the conceptual design.

Conceptual cost estimating has remained an art rather than a science in spite of all efforts injected over the years to come up with the final scientific solution. Although there is no likelihood that it can be subjected to precise mathematical treatment, methods have been developed that can provide more accurate estimates in less time. Such methods fall into categories which progress from a very rough "guesstimate" to a highly detailed procedure based on finished plans and specifications.

At this point it may be adequate to point out the similarity between conceptual cost estimating and approximate estimating. Many times these concepts are used synonymously. However, conceptual estimating aims at ascertaining costs before many of the decisions are made relating to the design of a structure. It should be structred in such a way that the consequences of various courses of action can be identified. Only then can it form the basis for cost control utilizing tools such as value engineering. The objective of an approximative estimate is only to provide a preview of the probable bid price, thus playing a largely "passive" role. Although the purposes for the two types of estimates may differ the techniques utlized are often the same.

Historically, estimating techniques might be divided into three categories (Logcher, 1980):

- productivity analysis estimates
- unit cost estimates
- parameter estimates.

Estimates of the two first categories require that components and their quantities are taken from the current status of the design. The cost per unit is then predicted from either historical unit costs for similar earlier components or from historical rates of resource usage for similar operations times the unit resource costs. Cost data or production rates are then adjusted to account for factors such as local conditions, time of construction or special project characteristics.In cases where sufficient detail about a design to meaningfully define its components and quantities is not known parametric estimating has to be applied. Parametric methods are characterized by the fact that the project cost is related to one or several "parameters". The parameters are chosen in such a way that historical data can be grouped to provide adequate input to the estimate. By decreasing the level of detail an increased sample size can be utilized.

Parametric Estimating

The purpose of parametric estimating is to produce a set of mathematical expressions that reflects the project costs using a minimum of data. The obvious most simple mathematical expressions relate the cost of a particular type of building to the size of that building, measured by its area or volyme. Data to support such relationships are published by several organizations such as McGraw-Hill's Dodge Divisions. Other simple parameters are of a functional nature. The cost of a hospital may be expressed as a function of the number of beds, the cost of a school as a funtion of the number of rooms and the square feet per student, the cost of a power generating plant as a function of plant-capacity, and so on.

It is obvious that the accuracy of such a rough estimate is low, probably too low to accept even in early planning stages of a riskful project. To increase the accuracy a number of parametric methods based on more elaborate mathematics have been proposed. In such methods cost estimating relationships (CER) are functional models that mathematically describe the cost of an item or activity as a function of one or more independent variables. The cost function can be defined by a relation

$$C = f\ (V_1, \ldots\ldots V_n)$$

in which C = cost of the building; and V_i = independent variables that specify building characteristics, such as location, function, height, structural type, quality etc.

Parametric estimating was first used in the US Air Force. It was used for cost evaluations in the development of new air crafts. The technique was later spread to other applications, and today some commercial systems based on the technique exist. One of the better known of these systems is PRICE by RCA (1978).

Kouskoulas and Koehn (1974) developed a cost function based on six variables specifying building locality, price index, building type, building height, building quality and building technology. Given these conditions for a proposed building it is possible to find a cost in dollars per square foot, and to comment with qualifications on the probability that the difference of the estimated cost from the actual cost will be within a certain percentage of the actual cost.

Parametric estimating techniques require that a complete record of the sample data is maintained with a concise definition of the variables and their values used in deriving the cost function. In estimating the cost of new projects it is instructive that the new project is compared with similar projects in the data for consistency in the assignment of the variables. It is also instructive that data is updated and the cost function is rederived periodically.

A major problem with parametric estimating refers to the control function of the estimate. There is no tie-in between such estimates and the detailed estimates (Logcher, 1980). During the period in the project development process when decisions with the most major cost impacts are made, adequate cost or cost change data relating to individual decisions are not available.

"Synthetic" Detailed Estimates

It is often necessary to develop cost data to compare the economies of various design approaches. Methods have been developed that "models" or "simulates" the detailed cost estimate usually made after the final design is determined. Such methods are based on the building component concept.

All types of buildings can be divided into the same building components, and each building component represents a function. The "structure" component's function is to frame the building, the "partitions" function to divide it, etc.

Usually a quantity takeoff procedure is required to identify all components and their quantities. By developing a quantity database from completed projects, the data in which is related to gross square foot of functional building space, "synthetic" quantity takeoffs can be generated from limited information about the design. The only information needed in some of the methods developed based on this idea is the total gross area of the building, number of floors, type of frame etc. Most building components can be evaluated on this basis.

In pricing the developed quantities up-to-date cost data is required. Such data is more exposed to changes than is the quantity data, and is therefore usually not stored in the database model. With an estimating approach in which quantities are generated from a historical database and the cost input is up-to-date data from the industry, it is possible to price any project at any location fast and in such a way that different design solutions can be evaluated. The estimate will closely resemble the detailed estimate generated from the actual drawings and specifications of the project.

Several models based on the above principle have been presented. The Building Optimization Program (BOP) developed by Skidmore, Owings & Mecrill is one of the oldest. Another commercially available model is Arches by the ICARUS Corporation.

One of the more advanced models is COSTMOD developed at MIT by Logcher (1980). Similar approaches have also been suggested elsewhere in Europe. In the COSTMOD model a hierarchical breakdown or tree representation of the project is used, where the project is the trunk and the components are branches and, finally leaves.

A model can be built for a class of projects and used by estimators to transform it into unique project trees for cost estimating. Project trees can be expanded based on the level of design decisions developed. Therefore, the system can be used throughout the duration of a project development.

The "synthetic" principle applied as in the COSTMOD and other approaches provides the capability of expanding data by simulating portions of the design process. Beginning with some basic information, a model can be developed to duplicate the designer's design and decision-making process. This type of model restates design experience and decision rules in mathematical terms. Their terms can vary in complexity, from simple rules of thumb to complex mathematical functions. As an example, open space area in

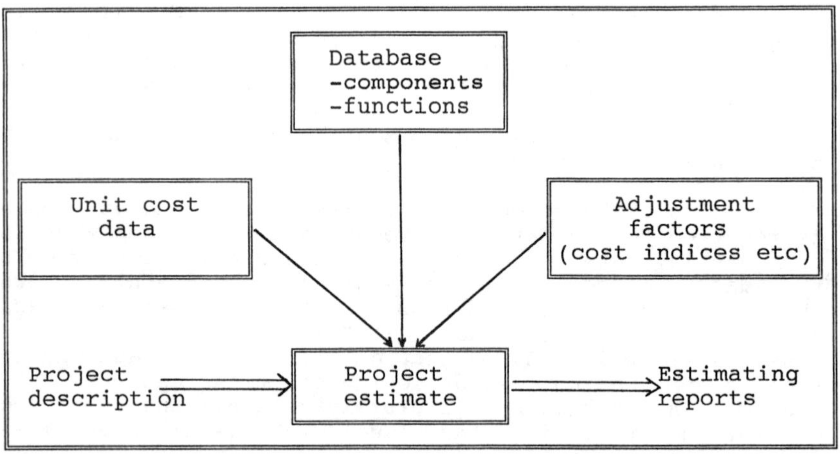

Figure 2. Overview of "synthetic" cost estimating

a housing project can be calculated as a function of the number of dwelling units, heights of buildings etc. Parameters of the function may be estimated from statistical analysis.

"Synthetic" cost estimating is a modelling process that requires a computer. The general philosophy of the method is shown in figure 2.

From the project description, which may be vague, the method can be used to forecast what are the detailed components of the project and their costs. As the design process progresses this forecast can be refined.

CONCLUSION

Two principles for conceptual cost estimating have been discussed. The first of these uses different parameters as references for the costs of a project without seeking to trace these costs to individual design elements. Such estimates have limited use as control systems. This value rather lies in their capability to fast generate approximative estimates of the costs.

The other approach simulates the final design and the design process yielding as result descriptions of components and their quantities. By applying relevant cost data to these quantities a detailed estimate is arrived at for the "synthetic" project. It is possible, even likely, that the final design will vary from the simulated one. But during the entire design process alternative design decisions can be evaluated.

Appendix

References

Kaufmann, G.A., Manager, PRICE Systems Applications. RCA/PRICE Systems, March 1978.

Kharbanda O.P., Stallworthy, E.A., & Williams, L.F., Project Cost Control in Action, Gower Press, 1980.

Kouskoulas, V., Koehn E., Predesign Cost-Estimation Function for Building, Journal of the Construction Division, ASCE, Vol. 100, CO4, December, 1974.

Logcher R.D., Techniques for Forecasting Costs, ASCE Specialty Conference on Building Economics, New Hampshire, 1980.

Seely, I.H., Building Economics, MacMillan Press Ltd., Bath, 1972.

Adding Knowledge Based Systems Technology to Project Control Systems

Robert D. Logcher, F. ASCE (1)

Introduction

Computers have been actively used in civil engineering design and construction for over 30 years. They have promoted improvement in productivity and effectiveness of the construction product. It is easy to point to areas of success, such as financial project control systems, network scheduling, accurate and detailed analysis procedures, automated component design procedures, etc. While these applications have changed our "way of doing business", they have also created a new set of problems, or shall we say, opportunities. When we have solved these problems, we will have created a new generation of project control systems which will provide far more effective tools for our industry. This paper will present a series of problems and describe how emerging new technologies can be used for their solutions.

The principal new technologies with which this paper deals all come from the disciplines of Artificial Intelligence. They include Knowledge Based (Expert) Systems (KBS), knowledge representation, object oriented programming, and natural language interpretation. These tools, not currently utilized in our project control systems, provide opportunities for drastically altering the character of these systems.

The computer has had some very deleterious impacts on our profession. In the same way as numerically controlled machines have de-skilled the machine tool industry, computers are de-skilling engineering design and management jobs. Already we see little need for a deep understanding of structural analysis and component design in the engineering office. These tasks have been automated. All the more reason why we need the next generation of tools, tools which can retain and exercise the knowledge no longer required of the designer. We must now deal with the analogies in design and construction management.

(1) Professor of Civil Engineering, Massachusetts Institute of Technology, Cambridge, MA 02139.

Early Capture of Information

This author has a bone to pick with much of the software industry which produces control systems for our industry. Most of these systems are generic and have evolved from financial accounting systems. In attempting to remain generic and deal only minimally with non-financial information, these systems have constrained their utility to historical record-keeping with little in the way of early warning or forecasting of problems. This leads to slow and late recognition of problems.

The issue revolves around both the early capture of data in a database so that it can be used for forecasting purposes and the breadth of scope of that data. It is well exemplified by using commitments from purchase orders and subcontracts to forecast cost at completion, variance from budget, and remaining exposure in job cost reports. Many systems don't use such commitment data. Also, payment requisitions forms, when sent to the field for data capture and entered into the computer, can provide immediate progress data while producing the invoice for the field. Input of a rough estimate with the identification of the need for a change order tracks both this need for the change order and its financial impacts.

The solution to this problem is embodied in more integrated systems, systems coordinated with better information flow procedures within companies, and a broader view of information content in the project control process. It is this latter solution component, a broader view of information, that is required if we are to utilize the new technologies mentioned above. For example, the mention of the need for a change order on a job, having its inception suggested by a particular subcontractor, may suggest a host of financial and management problems to the experienced project manager. Our future systems will attempt to capture this experience and mirror the reasoning of the project manager.

New Technologies

In this section, the technologies are mentioned and very briefly explained. They have not been invented by civil engineers, so we will not let them dominate this paper. Rather, our contributions lie in their application to our problems. Herein lies our challenge. Yet, it is worth mentioning that the character of our problems often challenges such technologies. Our problems tend to be larger, involving more disciplines and interactions, then the financial, mechanical engineering, and even VLSI industries. We tend to have larger databases, larger KBS's, more detailed CAD drawings and design data, and so on. As such, we will constantly be pushing these technologies.

The first technology in knowledge based systems. Fenves [Fenves, 1986] defines such systems by a separation of a knowledge base from a control or inferencing scheme which uses the knowledge, and an ability of a system to explain how the knowledge was used to reach conclusions. In these systems, domain dependent knowledge is usually stored in rules which should be understandable to the domain expert. (Note: the author uses the word "expert" guardedly, since real expert behavior is hard to corroborate, leading to more modest goals for most systems, and many systems now under development use multiple knowledge sources, including all of their users. In such cases, maybe we should call them apprentice systems.) With the separation of the knowledge base, these systems are easily changed to allow incremental growth.

Problem solving in these systems use a combination of forward and backward chaining through rules. In forward chaining, the rule base is checked using problem data for rules for which the premise is true. The action or conclusion parts of the true rules provide new data which might cause other rules to become true. When no more rules fire, the process ends and the data contains the solution. With backward chaining, a hypothesis of the solution is first generated using forward chaining, and then attempts are made to verify the hypothesis by looking at rules which contain the hypothesis in their conclusion and seeing if their premise are true. If all data in their premise is not known, the process chains backward by attempting to determine such data in the same manner. If the backward chaining finds data to verify all premises in the chain, the hypothesis is true, and if not, false. Then another hypothesis must be generated and tested.

Pure rule based systems are only useful for small problems where all the knowledge can be represented in several hundred rules. The problem with rule based systems is that their efficiency degrades exponentially with knowledge base size. [As shown in Niwa, 1984] As the breadth of knowledge increases, the premise of the rules become longer and more complex in order to apply the rules to their appropriate subset of the problem domain. Checking for applicable rules also takes longer as the knowledge base size increases. Since we expect our systems to become integrated and very large, this form of system is will not work acceptibly. We must therefore look toward the concepts of object-oriented programming [Abelson and Sussman, 1985] and frame-based systems. A frame or object is analogous to a record in a database, with numerous additional characteristics. One object may be a pattern for its instances, where the instance objects then inherit its slots, or knowledge holders, and slot characteristics. Slots may hold data, and slot characteristics might include a data verification procedure, of which acceptible bounds is the simplest example. Slots may also be programs or procedures or knowledge such as rules or rule bases. Objects may be related in more complex manners analogous to the set relationships in network databases. The concept of inheritance may then be associated with any relationships so that objects can take on properties of several objects to which they are related. An activity can inherit an algorithm for calculating its early start and finish dates given those of its predecessors from a generic CPM activity. It could also inherit knowledge about how to figure its duration from knowledge about its type of work. Finally, slot characteristics may include procedure or active rule bases. Then, when a data value in a slot is changed, this automatically triggers actions.

The concept of message passing is central to object oriented programming. One object, operating independently, can reach some conclusion and then inform other objects to which it is related of this conclusion. Messages sent to an object cause it to store data and initiate procedures which check the impact of the message on itself. This may result in changes within itself, changes which may in turn generate messages to other objects. A blackboard, or message and data coordinator, is often used to control reasoning processes. The section after the next in this paper provides a detailed example of the use of these concepts.

The application of such features is quickly apparent. Rule bases can now be disaggregated, leading to efficiency and disaggregate collection of knowledge. Procedural knowledge can be conveniently mixed with other forms. If we think of a building design stored in this way, when a specified pump is unavailable during construction, the field personnel provide a table of available pumps, the pump object redesigns itself and then signals its interfacing technologies, electrical and structural, of the interface changes, which are then checked and, if necessary, changed or their designers signalled of a required change. Change then propogates, providing change management.

The frame based structure provides an effective tool for planning paradigms. Scenarios of conditions, actions, and their effectiveness could be stored as an historical knowledge base. When a planning problem is recognized and immediate knowledge not sufficient for a solution, the knowledge base is searched for similar conditions and alternative actions thereby generated for further screening. This structure may be more efficient then the generate and prune algorithms using a single fixed knowledge base.

Lastly, we can apply these techniques to create intelligent database systems. Such systems have daemons which the system scheduler starts on a periodic basis. They traverse the database, looking for changes in data since their last pass. When finding changes, a daemon would start an inferencing process to check for impacts of the change and look for patterns of changes and causes. This might then initiate forecasting and planning processes, thus leading to much more dynamic project control systems.

Natural language translation is obviously a technology closely linked to all applications. Its use is for more natural, user-friendly communication with these systems, to provide knowledge capture, data input, and processing control. It application together with KBS's is most interesting. While numerous natural language query systems are available, often closely related to database managers, such as CLOUT with R:Base or INTELLECT with FOCUS, their domain is realistically limited to direct retrieval and minor manipulation of data from one or more files. Dictionary words are associated with individual fields or groups of fields in a file or very simple direct operations on the fields. New systems, such as Expert-MCA being developed at M.I.T., use deeper user-defined knowledge of the meaning of data to answer more complex questions. [Logcher, 1986] Such knowledge defines pattern searches against the database.

Database Data Analysis

When we look at our project control process, we are struck with the realization that the computer tools we use are so predominately generic, having almost no knowledge of project, company, or construction technology, that we must introduce very large amounts of manual processing to gain information out of our systems. Our CPM schedules are generated for every project from scratch, or maybe from a previous similar project. Our job cost budget may be generated from our estimate, which was similarly generated. We are starting to see some estimating systems [QuickEst, 1985] which use cost modeling techniques, where a component model does incorporate design and construction knowledge. This section and those that follow try to provide examples of how such knowledge can change our project control process. The next section shows how the technology of object oriented programming can help in the implementation.

With our database technology, we have been able to collect, process and report very large amounts of project data very easily. As a result, we have made decisions to disaggregate our project down to the smallest responsible party. We can then report accomplishments and compare against expectations for all parts of our project. We then overwhelm users with data. (Note: the author does not say "information".) Exception reporting might keep down the volume of results, but does little to help us understand the causes of our problems. This is still a hard, knowledge intensive process, hampered by having a narrow window into our project in which updating takes place (small percentage of accounts).

IPMS [Navinchandra and Logcher, 1986] attacked this problem. Taking a typical job cost report with responsibility for an account shared between estimator, superintendent, and foreman, it showed how a rule base could be used to search for patterns among accounts to evaluate performance. Data needs to be conditioned to eliminate estimator bias before field personnel can be evaluated. Determining whether the poor performance of one foreman was the fault of the foreman was found by looking at the performance of other foremen working under the same superintendent. A pattern of poor performance indicated that the superintendent was most likely at fault. Figure 1 shows a typical rule in this system, while Figure 2 shows some typical results. An explanation capability is included.

Example Application of Object Oriented Programming

It is worth illustrating the character of this technology through a project control example. While the example, calculation of dates in a CPM schedule, could easily be solved with a simple algorithm dealing with all the network data together, the application of the concept of disaggregated knowledge become clearer. This application has been implemented by the author and his students in KEE [IntelliCorp, 1986] on an AI workstation.

In its simplest form, the system contains generic objects called PROJECT, ACTIVITY, and RELATION. Each of these contain slots for the data typically stored in algorithmic programs. In addition, these objects contain some very simple procedures, each consisting of small steps from the algorithmic solution process. When we solve the scheduling process with pure message passing [Abelson and Sussman, 1985] we will store some data in the objects which are duplicates of data in other objects, but we will show the value of this duplication.

```
Rule:    (for CaseIX)

  IF     <the account has ACOST reported>
  AND    <the AQTY is reported>
  AND    <the REVQTY is reported>
  AND    <the ETCCOM is not reported>

  THEN   <it may be concluded the the account is
          a CaseIX account>


Rule:

  IF     <an account has high actual costs>
  AND    <the ESTIMATOR has a history of low estimates>
  AND    <the SUPERVISOR has a good history>
  AND    <the FOREMAN has at least a moderately good
          history>

  THEN   <it may be concluded that the ESTIMATOR is at
          fault>
```

Figure 1. Typical IPMS Rules.

The foreman David Brown shows a tendency to overspend by 30 % while working for super Tom Fulton, who also has a tendency to overspend

Estimator Mark Wilson underestimates by 12.5%

Foreman David Brown overspent on ACCT 35, but did OK because estimate was too low

Figure 2. Typical IPMS Results.

The generic PROJECT contains a procedure called CREATE.PROJECT. Whenever this procedure is initiated (by sending it a message), it will prompt the user for project data, name, start date, finish constraint, etc., create a project instance object called PROJECT.name, and then allow the input of activities and relationships. The activities are instances of the generic activity and are related to the project instance. Data stored in the activities and relationships are durations, lead/lags, and network topology.

The algorithm is initiated by sending a message to the INITIAL.SCHEDULING procedure in the object PROJECT.name. The procedure is not stored in the object, but inherited from the generic PROJECT. This procedure sends messages to each activity telling each to initiate itself (set early and late dates to NIL, erase any dates from predecessors and successors). This message contains the project start and finish dates. For simplicity, we will assume that the project has a specified finish date. Activities receiving this message can handle it without needing information from other objects. If an activity has no predecessors, it can take the project start date and schedule itself. It then send messages to successor relationships informing them of its schedule. The relationships in turn send messages to their successor activities. The activity, receiving a message from a relationship, stores the message data and decides if all predecessors have reported. If so, the activity can be scheduled and the process propagates itself to the terminal activities. If an activity has no successors, it uses the project finish date in the same manner and proceeds with the backward pass at the same time as the forward pass. Float is calculated by whichever pass processes the activity last.

While this might seems like a complex process, it is simple and efficient. The procedures are small and inherited from the generic objects, not duplicated in each object. The real benifit comes when we realize that with this same structure, we can maintain a schedule during progress reporting. With only slight changes to the activity procedures, the impacts of progress reports can be propagated forward and backward only as far as they change schedule dates. We are using this with a knowledge base which analyzes project data (Note: not CPM data) such as charges to cost accounts or down time for a particular piece of equipment to send messages to activities about actual starts, finishes, changes in duration, etc. From these changes, recalculation is automatic.

To extend this concept further, we might consider the procedure for calculating the early dates for a activity. The procedure in the generic activity is simply to add the duration to the start date if a finish date is not determined from a start or finish to finish relationship. We could, however, substitute for this generic procedure a smarter procedure that understood that the productivity of the particular activity was weather sensitive and that the duration or finish calculation should be a function of the time of year of the activity start. This procedure could be inherited based on the type of activity.

The point of this example is that this technology allows us to insert easliy into the control systems knowledge and relationships about our data and use it directly in forecasting and problem detection.

CPM Scheduling

It is easy to see that existing CPM systems do not solve our real project management problems. Problems abound, from the tremendous amount of knowledge that must be applied to generation, updating, and interpretation of schedules to the well-known time/cost creep problem. The development of KBS's using object oriented programming techniques are geared solving these problems. These techniques have been applied recently in various research efforts. [Arias-Toro, 1986, Levitt, 1985, Hendrickson, 1986] The problem can be decomposed into several parts, each needing different inferencing techniques. These parts are:

1. Schedule generation - task identification, task design, and sequencing

2. Progress reporting - automatic schedule updating from database data

3. Analysis of variance - determining the cause for performance outside of expectations

4. Projection of variance onto remainder of schedule

5. Replanning to overcome impacts and mitigate causes of variance

Schedule generation is a planning process which can use many of the planning paradigms. Arias-Toro developed a knowledge structure which ties project characteristics to subnetworks with rule bases. A rule base is fired by the existence of a characteristic and is used to set subnet activity characteristics and connect the subnet into the project network. Physical, geometric, and construction technology constraints on the interactions between activities [Logcher, 1987] can also be used to generate precedences. These works also show how hierarchical detailing may be used to increase the detail in schedule design when needed to achieve schedule goals.

Currently, progress reporting involves manual input of activity starts, finishes, percent complete, and remaining durations. This data is abstracted from the broad information available about the project. Using the concepts of intelligent databases and domain knowledge in the daemons, this process will be automated. The more difficult problem is determining the causes for variance. A preliminary knowledge structure for this analysis is shown in Figure 3. [Nay and Logcher, 1985 and 1986] Knowledge of potential causes of variance must be precoded, with rules embodying why an activity or work package might be sensitive to the cause. When a project is designed, each work package can be analyzed for its sensitivities. When variances are noted, these sensitivities represent first hypotheses for causes. Hypotheses are varified or refuted by checking other work packages with similar sensitivities as well as communicating with the project manager for outside influences and his ideas.

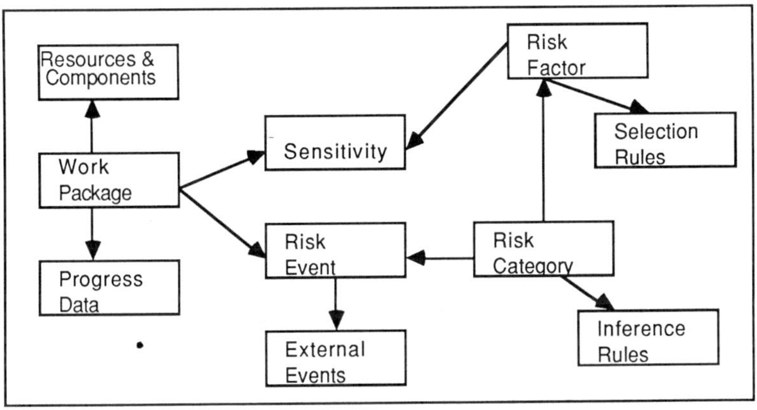

Figure 3. Knowledge Structure for Analysis of Variance.

Conclusions

The applications mentioned here are only a start. Financial control, while not discussed specifically, has a direct analogy with schedule control. Throughout the discussion, it is apparent that a separation is inappropriate and unnecessary. Our future systems will be far more integrated and deal in a common manner with all types of project, environment, organizational and other data, using knowledge to integrate and utilize the data.

References

Abelson and Sussman [1985] Abelson, Harold and Gerald J. Sussman, Structure and Interpretation of Computer Programs, M.I.T. Press, Cambridge, MA, 1985.

Arias-Toro [1986] Arias-Toro, Mauricio, Knowledge Based Planning Schedules for Public Works Projects, unpublished S.M. thesis, Department of Civil Engineering, M.I.T., Cambridge, MA, May 1986.

Fenves [1986] Fenves, S. J. "What is An Expert System," in Expert Systems in Civil Engineering, Proceedings of the ASCE Technical Council on Computer Practices, Seattle, WA, April 8-9, 1986, ASCE, New York, 1986.

Hendrickson [1986] Hendrickson, Chris, Carlos Zozaya-Gorostiza, Dan Rehak, Eduardo Baracco-Miller, and Peter Lim, An Expert System Architecture for Construction Planning, Engineering Design Research Center Report EDRC-12-07-87, Carnegie Mellon University, Pittsburg, PA, December 1986.

IntelliCorp [1986] KEE Software Development System User's Manual, KEE Version 3.0, IntelliCorp, Mountain View, CA, 1986.

Levitt [1985] Levitt, Raymond E. and John C. Kunz, "Using Knowledge of Construction and Project Management for Automated Schedule Updating," Project Management Journal, Vol. XVI, No. 5, pp. 57-76, December 1985.

Logcher [1986] Logcher, Robert D., Ayman M. Hindy, and Ming-Teh Wang, ESCUPAD: Initial Knowledge Based System for Processing Queries to CAPCES, Research Report R-86-10, Order No. CCRE-86-3, Department of Civil Engineering, M.I.T., Cambridge, MA, 1986.

Logcher [1987] Logcher, Robert D., D. Navinchandra, and Duvurru Sriram, "GHOST: A Project Schedule Generator," submitted for Expert Systems in Civil Engineering, J. of Technical Council on Computer Practices, ASCE.

Navinchandra and Logcher [1986] Navinchandra, D. and Robert D. Logcher, "Steps towards Automatic Interpretation of Project Data: A Knowledge Based Systems Approach," in Microcomputers in Engineering, B. A. Schrefler and R. W. Lewis, ed., Pineridge Press, Swansea, U.K., 1986.

Nay and Logcher [1985] Nay III, Leston B. and Robert D. Logcher, An Expert System Framework for Analyzing Construction Project Risks, Research Report R85-4, Order No. CCRE-85-2, Department of Civil Engineering, M.I.T., Cambridge, MA, 1985.

Nay and Logcher [1986] Nay III, Leston B. and Robert D. Logcher, "Proposed Operation of an Expert System for Analyzing Construction Project Risks," Proceedings of the Nineth Conference on Electronic Computation, Structural Division, ASCE, February 23-26, 1986, pp. 65-76.

Niwa [1984] Niwa, Kiyoshi, Koji Sasaki, and Hirokazu Ihara, "An Experimental Comparison of Knowledge Representation Schemes," in AI Magazine, American Association for Artificial Intelligence, Vol. 5, No. 2, pp 29-36.

QuickEst [1985] QuickEst Version 2.00 Estimating System, Program Description/Operations Manual, Constructive Computing, Kansas City, KS, 1985.

PROJECT CONTROL SYSTEM INTEGRATION

By

A. A. Badger
Senior Project Manager
Stone & Webster Engineering Corporation

Dr. K. A. Reinschmidt
Vice President
Stone & Webster Engineering Corporation

A. Richard Gandt
IBM Corporation
Waltham, MA

ABSTRACT

This paper describes current project control systems developed over the past two decades and utilized by Stone & Webster Engineering Corporation to control large international and domestic projects.

It further describes how these systems are being improved based on recent advances in computer hardware and software to be more integrated with the overall project engineering, design, construction, and operations processes.

In essence, it describes existing project control systems and procedures and identifies steps Stone & Webster is taking to apply the latest computer technology to improve cost and schedule control and performance.

INTRODUCTION

Project Management, utilizing currently available computer hardware and software, is on the verge of a breakthrough equally as dramatic and challenging as introduction of critical path scheduling. Technological development enables computer-based management control and technical systems to be an integral part of management, rather than supplemental information systems.

The past two decades saw the development and widespread introduction of Project Management Systems utilizing Critical Path Method (CPM) calculations for schedule analysis, work breakdown structures for scope definition and assignment of responsibilities, and resource allocation for evaluation of manpower requirements. Project Control Systems, however, were thought of by many as an accessory which monitored the engineering/construction process. [They were not truly integrated into the process until recent hardware/software developments made complete integration possible.]

Similarly, in Engineering and Design, CAD and CAE programs enhanced the ability to create and store two-dimensional drawings, but plastic models and separate data systems were still required for interference detection, equipment lists, engineering analysis, configuration control, and project control.

Current three-dimensional (3-D) computer graphics combined with advanced data base technology have provided a new and powerful tool to automate and integrate the project management process with the project technical process.

This paper addresses how many of these traditional discipline oriented functions are affected and the new opportunities provided for managing complex projects.

INTEGRATED MANAGEMENT SYSTEMS

Complex projects require comprehensive management systems to effectively manage the overall project from engineering through procurement, construction, and commissioning. Over the past 15 years, Stone & Webster Engineering Corporation has developed and implemented a very effective management system called the Stone & Webster Integrated Management System (SWIMS) on all major projects. The system provides a systematic methodology for:

- Defining the scope of the project
- Defining responsibility for carrying out various aspects of the project
- Developing realistic schedules and cost estimates
- Controlling materials
- Staffing to meet schedules and man-hour estimates
- Measuring performance

The Project Manager and his staff are responsible for implementing the project control methodology across all project participants. This planning process begins at the earliest possible time, utilizing a Work Breakdown Structure (WBS) preferably before project authorization.

Immediately after project authorization, the planning process (Figure 1) intensifies by means of an intensive planning session involving project supervisory and management personnel, including, if possible, the client's representatives. During these sessions, the work breakdown structure is further refined. The assigned project team under the direction of the project manager further breaks down the project objectives into manageable work packages which are utilized to control the project in terms of specific end products for which:

- Costs can be estimated
- Activities can be scheduled
- Responsibilities for accomplishment can be assigned
- Costs can be collected

Each work package has a written definition of scope which is a communications tool for project participants and can later be used as a basis for change control. The integrated project objectives, resource loaded schedule, and budget are defined as the project base plan.

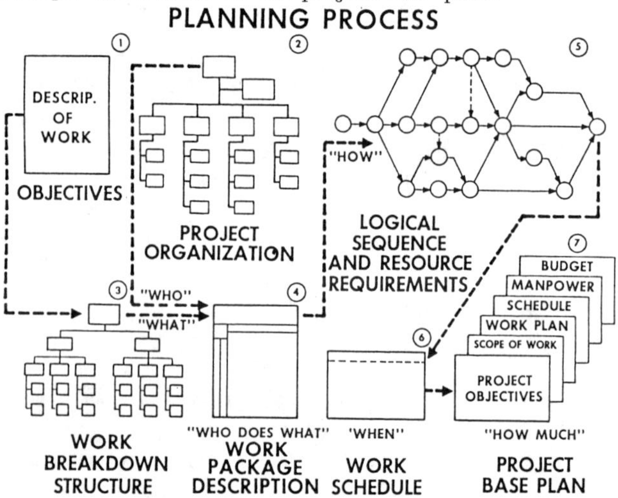

FIG. 1

Construction Management System (CMS)

The Construction Management System (Figure 3) originally developed in the early 1970s is a subsystem of SWIMS which may be operated independently or in conjunction with the Engineering Management System to control the entire project.

The first step in construction planning is to break the project down into logical management areas utilizing the project plot plan and building layouts. The management areas are then further subdivided into work packages considering the requirements of engineering material control and construction management.

Construction Work Breakdown Structure

The construction work breakdown structure is the basis for:

- Defining project scope
- Establishing project schedules
- Assigning resources
- Collecting progress
- Summarizing reports for various levels of management

The work breakdown structure (Figure 2) is further utilized to define the account structure for estimating quantities of materials in each management area. Man-hours are estimated utilizing appropriate man-hour unit rates which are adjusted based on the expected productivity of the domestic or foreign expatriate contractors and prevailing working conditions. CMS utilizes the quantity and unit rate data to automatically manload the networks and produce the basic productivity reports and required resource curves. The system is then utilized to measure productivity

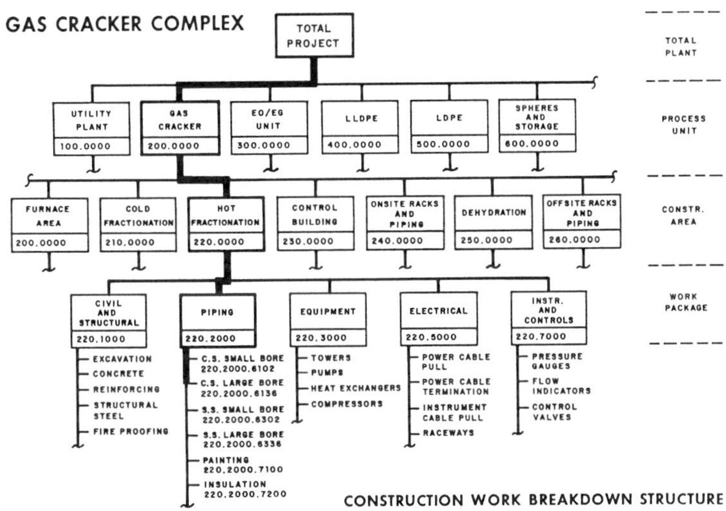

FIG. 2

of the workforce by collecting actual quantities in place against each network activity and comparing the man-hours required to complete the work to the actual man-hours expended.

Cost & Schedule Performance is measured utilizing both CPM and earned value approaches. The earned value consists of establishing a Budgeted Cost of Work Performed (BCWP) for each item of quantity assigned to a network activity.

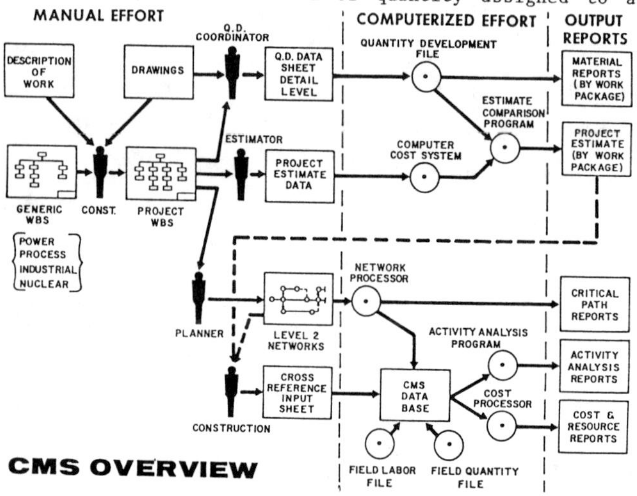

FIG. 3

Manloading curves are also produced by craft, management area, or total project for evaluation of resource requirements, congestion, required rate of buildup, peaking and eventual force reduction based on available work and the schedule logic. Simulations are easily accomplished by adjusting various aspects of the schedule.

Material Management

An extremely important aspect of managing projects, especially an international project, is control of materials. This consists of controlling both bulk materials, mostly pipe, structural steel cable and instruments, and tagged or engineered equipment such as tanks, heat exchangers, pumps, etc. Stone & Webster utilizes a bulk material system which interfaces with both EMS and CMS to control bulk materials and other listing programs to control tagged equipment.

The material system assists in procuring piping and other bulk materials materials by issuing requisitions for materials based on separate manual takeoffs. The preliminary takeoff consists of an estimate of the required materials based on takeoffs from piping and instrument diagrams (P&IDs) and other preliminary drawings by CMS management area. The second or intermediate takeoff is performed sometime later when planning or production drawings are available in each management area. Final takeoffs are performed as isometric and final construction drawings are completed.

As each takeoff is made, the quantity of material is compared to the previously requisitioned material file and final requisitions are issued for any materials not found in the file.

One of the most important functions on an international fast track project is an accurate material takeoff (MTO) done rapidly as drawings are developed. Every nut, bolt, gasket, and fitting must be accurately identified with appropriate contingencies for loss, damage, etc, and promptly requisitioned consistent with the construction need date.

For this to work effectively, the Engineering Management System (EMS) must accurately schedule when drawings will be issued for each management area and when the MTO will be performed in the home office. Adequate lead time must be provided to requisition, fabricate, and deliver long lead items such as large fittings or heavy wall pipe. Unique materials requiring special fabrication steps such as lining and/or coating must also be identified, allowing for adequate lead time. And, finally, an accurate inventory control system is required to establish when materials are scheduled to be shipped, and are actually shipped, received, and warehoused.

Engineering Management System (EMS)

The engineering management system (EMS) has essentially the same components as the CMS system, namely:

- Work breakdown structure
- Critical path network
- Man-hour estimates
- Progress reports

The engineering work breakdown structure (WBS) is similar to the construction work breakdown structure except that the work packages differ somewhat in scope. The products listed for each work package are essentially engineering products rather than construction end products. The engineering WBS is utilized in the same manner as the construction WBS in that it is used:

- To break down the project into manageable work packages
- To estimate home office man-hours based on the scope of work
- To develop work package CPM networks to schedule when the work will be performed
- To collect actual man-hours expended by work package to evaluate performance
- To produce management reports at various levels
- As a basis for controlling changes

Numerous reports are available from EMS to control the project cost, schedule, and resource availability. Performance in EMS is measured similar to

CMS; however, physical quantities are not measured in engineering as they are in construction, and the BCWP is based on the man-hours estimated in EMS to complete an activity.

Various graphs and cost analysis reports are produced at the work package level up to the total project level to assist management in assessing the manpower required by discipline, and the effectiveness of that manpower in meeting the project cost and schedule objectives.

ENHANCEMENTS TO PRESENT SYSTEMS

The existing control systems have several aspects where major advances in computer application can enhance their effectiveness. These are:

- Better integration with the engineering design construction process
- Elimination of errors in the engineering design procurement process which require rework and disrupt the schedule
- Better visualization of the project with respect to:

  Construction sequence
  Equipment access
  Work accomplished
  Problem areas

- More effective use of manpower by reduction of manual efforts required to take information from one document such as a drawing and fill out another document such as a purchase requisition

INTEGRATION WITH THE ENGINEERING DESIGN PROCESS

Existing project management systems simulated the thought process through engineering, design, procurement, construction and startup. CPM networks are used to model the logical steps in the process and establish dates for each discrete step utilizing arrow or precedence diagrams and estimated time durations.

Once construction starts, the accuracy of the project system is primarily dependent on the time required to input and update the existing data base with new and more accurate information as the engineering and construction proceeds.

By utilizing state-of-the-art computer technology, Stone & Webster has greatly improved on the accuracy wind flow of the information through a project.

Previous large projects depended on utilizing plastic models for:

- Elimination of interferences
- Visualizing equipment access for operation and maintenance
- Visualizing construction sequence and access

However, to build the plastic model, the design had to be finalized to a substantial degree. A long time period usually elapsed before drawings were accurate enough to build the model and the plastic model itself takes some time to build. The model sometimes then precipitated changes as construction and owners visualized the configurations and reviewed operation and maintenance procedures.

In addition, the design had to be well advanced to develop accurate quantities to validate the times estimated for construction activities.

In many cases the lead time between model construction and actual construction was insufficient to eliminate interferences, material quantity errors, or arrangement problems.

THREE-DIMENSIONAL SOLIDS MODELING

To improve the entire project development process, Stone & Webster has developed an integrated approach to computerized plant modeling for engineering, design, construction, operations, and maintenance. This system runs entirely on IBM computers, and the data base developed by SWEC during the design and construction process can be used during the plants entire life. The major elements of this system are as follows:

- A mainframe computer for central storage of the integrated plant database, including drawings, diagrams, geometric data, and operational data.
- Color graphic workstations for plant design and viewing of the three-dimensional color models of the plant
- Personal computer workstations for database access and viewing of drawings and models
- A three-dimensional solids modeling software package
- A two-dimensional drafting package
- A relational database system

In the Stone & Webster approach, selected portions of the plant are modeled in the computer (Figure 4) as three-dimensional solid objects before the two-dimensional drawings are prepared. This permits each piece of equipment to be individually selected and located in the plant. Pipes, cable trays, structural steel, and other connecting elements are also modeled and located as spatial components. Thus, changes and retrofits are easily and quickly accommodated, as individual components can be moved as entities. The computer program will also check for any interferences as the equipment is being located or modified.

Because the plant is modeled in true three dimensions, material quantities can be automatically taken off, and changes can be processed automatically. Construction drawings are made by creating slices or views of the appropriate models either horizontally (for plans), vertically (for sections), or obliquely (for isometrics).

The same computer models are used in constructibility review and planning. The movement of equipment can be planned in the true three-dimensional geometry, and cranes can be positioned in the best locations for major lifts, with no interferences. Complex three-dimensional shapes for reinforcing bars, concrete formwork, sheet metal, and other materials are developed and visualized more easily than in two-dimensional drawings. The construction sequence for the plant is developed and visualized in three dimensions by placing components sequentially. Through the database, the material quantities and construction sequence information is tied to the construction CPM networks and to construction work packages. The time required for construction planning is greatly reduced and design problems that would cause problems in construction are eliminated in the design phase.

The computer model has many additional advantages over a plastic model:

- The model always represents the present design, eliminating the time gap which occurs when building the model from drawings. This reduces rework on both the drawings and the model and eliminates repetitive material takeoffs and errors due to timing.

- The interface of the physical configuration with analysis programs such as NUPIPE and STRUDL improves the efficiency of engineering.

- The identity of a component as an object in the model allows us to relate technical schedule and cost data in the database to the model. The direct relationship of the physical configuration with the technical requirements instead of to reference documents is a new and powerful tool.

- The preliminary model can be viewed from many workstations and locations at the same time, by engineering, the owner, construction, operations, etc. This is a powerful communications tool to speed up decisions and resolve problems.

- The model can be used to simulate operations such as component installation and removal, radiation monitoring, security evaluations, erection sequencing, etc. This provides a powerful tool to try various options and achieve the optimum sequence. It also facilitates visibility by management, client, and supervisory personnel.

- A model can be created during the bid phase to establish sequences and schedules, and to quickly estimate construction methodology and material costs, reducing the cost and uncertainty in lump sum proposals.

- Plant operators can use the three-dimensional model to review the plant design for operability and maintainability, thereby finalizing the layout early. Operators can walk through the computer model of the plant and assess equipment removal areas, maintenance spaces, truck aisles, platforms and walkways, and other aspects of the plant while they are in the design stage.

- Feedback of as-built information from construction, provides a final geometric model for future maintenance and modification functions.

PLANT DIGITAL MODEL

One of the most essential aspects of project management is managing the flow of information from engineering and design through the appropriate project functional groups that manage the procurement and construction activities. This has been accomplished by implementation of a series of semi integrated; cost, schedule, and material control systems. These systems provide indications of potential problems which the Project Manager must detect and, if possible, provide corrective action. His and other project participants' ability to correct these problems depend on the timeliness and accuracy of the data upon which they base their decisions.

Integration of a three dimensional ("3D"), Plant Digital Model (PDM) into the entire project process greatly increases both the accuracy and timeliness of available information since the information is derived concurrently as a byproduct of the design process. This system working in "real time" not only eliminates the lag in feeding the most current information into the project control systems, but it eliminates many of the manual transactions which are, in effect, the source of delay and inaccuracy in the technical processes.

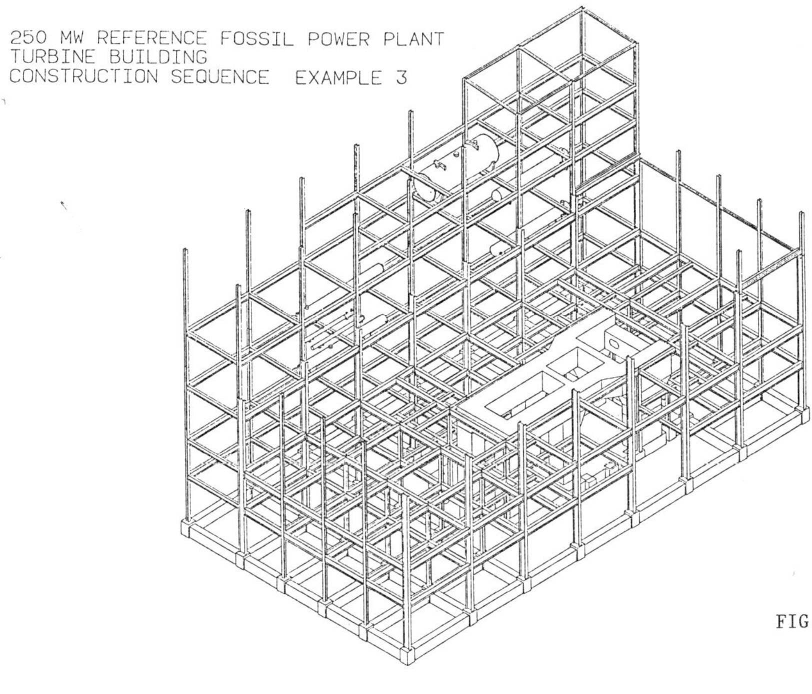

FIG. 4

INTEGRATION OF CMS WITH PDM

The following (Figure 5) shows the CMS system integrated with the PDM and a relational data base (IBM DB2).

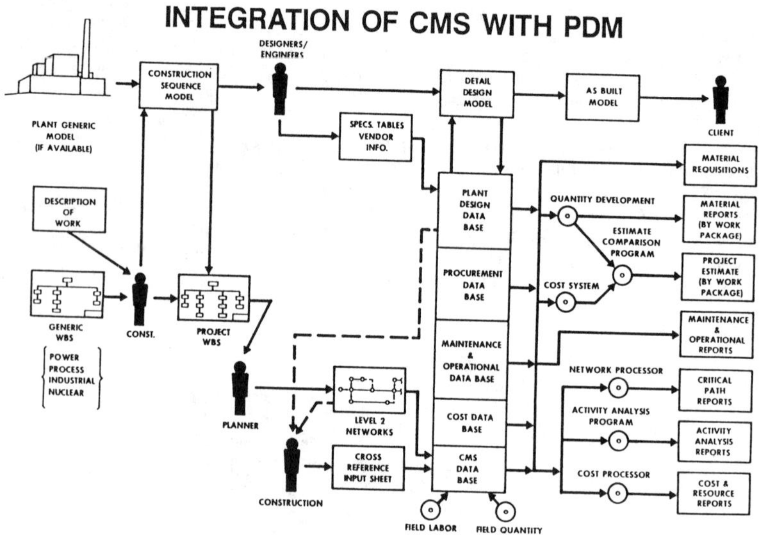

FIG. 5

In this scheme, many of the manual processes such as drawing production and material takeoff (MTO) are performed automatically by the computer accessing data directly from the "3D" model.

The process starts as before with development of a Work Breakdown Structure (WBS) based on the Description of Work and whatever other project information is available. At the same time, a construction sequence model is developed from the Description of Work and any other information which is available. These two entities (e.g., the construction sequence model and the WBS) will be the basis for defining the work packages to structure and control the information in the project database. The construction CPM network is also developed from the construction sequence model.

The PDM project database is greatly expanded from the previous limited CMS database, which contained only cost and schedule information for a new construction project. The expanded database is a relational database with original CMS data, plus (a) a plant design database (such as equipment and material sizes, configurations, specifications, locations, materials, vendor data, connections, etc.), (b) a procurement database containing procurement data (such as procurement specifications, vendor P.O. data, prices, order status, etc.) for the items in the plant design database, (c) maintenance database (maintenance requirements, procedures, spare part availability, in-service status, etc.) for all equipment and components requiring maintenance from the plant design database, (d) a cost database (unit man-hour rates, craft distribution cost per man-hour by craft, estimate category, etc.).

In addition, the same data is all stored only once eliminating redundant files and storage costs.

SYSTEM INITIATION

The system is initiated by developing the necessary technical data tables and projectizing the construction sequence model and the WBS.

Once the construction sequence model and the WBS have been projectized and the necessary tables of engineering data have been created, the detail design and project control process starts.

As the plant digital model (PDM) evolves, information is categorized and fed into the plant design database where it is available to all of the information systems used to plan and control the design, procurement, and construction processes.

As information is fed into the procurement database, judgments can be made by the project when to issue requisitions to vendors for bulk materials such as piping, cable tray, cable, valves, etc.

As vendor data is received from vendors for major equipment, it is also fed into the plant design database for updating the plant arrangement, final routing of piping, cable, HVAC, etc., and calculations may be performed utilizing the database information to confirm sizes of supports, nozzle loads, flows, temperatures, etc. An up-to-date status of what vendor data was used in any particular calculation and if it was preliminary or final.

Utilizing this information and vendor maintenance procedures, a maintenance and operational database is maintained for the plant. The database includes installation date, required preventive maintenance schedule and procedures, spare part availability, operating, or in-service time, test status, etc.

The cost database contains data to compute estimated costs of the work, actual costs of the work, forecasts to complete, records of changes in quantities, man-hour rates, etc.

The CMS database operates essentially as it had previously except it has available to it more accurate sources of quantity data on which to base the overall schedule and cost analysis projection.

As the actual quantitites are installed, they also are stored in the database so a version of the PDM can be created showing actual quantities in place. These can be overlayed with the construction sequence model and the schedule dates to show critical problems in installation of quantities scheduled vs. quantities ordered and delivered. Critical material delivery problems are identified well in advance and corrective action taken to correct any serious delivery problems.

CLIENT TURNOVER

Final completion of the project and turnover to the client also can include turnover of a fully functioning material and maintenance control system. This design basis database has stored in it all of the plant technical specifications, drawings, critical as-built dimensions, vendor data, maintenance and operations records, cost history, procurement history, etc. It is apparent that having this information available in a readily retrievable, paperless format will greatly enhance the value of this information. This feature alone would justify the cost of implementation over the life of the plant.

CONCLUSION

Project management has adopted the computer for over two decades to perform critical path calculations, to sort lists of data such as material reports, cost estimates, and for functions such as records management and retrieval. There are numerous project management computer packages now in the marketplace which will run on mainframes or microcomputers. With the exception of a limited number of features such as resource loading, hardware compatibility, enhanced graphics, and the vendors perception of "user friendliness," the programs are functionally very much alike and have not changed dramatically since the introduction of PERT/Cost. In essence, they monitor the project function and provide the Project Manager with a tool to "help or force" the project to plan and control its work with varying degrees of success.

The new more powerful computer hardware and graphic and database software which is now available not only improve the efficiency of the engineering, design, procurement, and construction process, but also provides a powerful management tool to integrate these functions.

AN IMPLEMENTATION STRATEGY FOR IMPROVING PROJECT CONTROL SYSTEMS

by

C. WILLIAM IBBS[1], DAVID B. ASHLEY[2], JAMES M. NEIL[3] AND FRANK W. FEILER[4], M. ASCE

ABSTRACT

Present-day project control systems, as a part of Computer-Aided Construction (CAC), have found wide use and value in engineering and construction practice. Part of this demand stems from the flexibility and power of the software packages themselves, but a very important reason also comes from the emergence of personal computers and their affordability. To those close to the Industry, it seems clear that personal computing facilities are: 1) here to stay, 2) surprisingly ingrained technology, and 3) still in need of some software and hardware improvements. This ASCE Specialty Conference, which this article summarizes, has examined the deficiencies and needed improvements from the user's perspective of the software. This particular paper attempts to cast some order to these "wish lists" and proposes a development strategy to achieve these changes.

INTRODUCTION

Project management control systems, which combine simple and complex algorithms, calculations, external file interfaces, database management and a host of other features, are a natural for computerization. The most primitive of systems, which have been available on the largest of machines for twenty years or more, allow the capture, analysis and reporting of data in a variety of formats. Even the earliest of packages allowed some degree of "what if" and "how

---

[1]Assoc. Prof., Dept. of Civil Engineering, University of Illinois, Urbana, IL 61801.

[2]Assoc. Prof., Dept. of Civil Engineering, University of Texas, Austin, TX 78712-1076.

[3]Manager, Project Controls, Power Group, Morrison-Knudsen Const. Co., Boise, ID 83700

[4]Vice-President, Rocketdyne Div., Rockwell International Corp., Canoga Park, CA 92304.

come" analyses, which after all is the function of project managers. To date though there has been a substantial gulf between what the software developers have provided and what the system users actually have sought. The preceding nine papers and the discussions that have surrounded their presentations represent one of the few attempts to bring these users and developers together in a thought-provoking, "no holds barred" forum with the express purpose of getting systems that really do help project engineers and managers.

In this paper we summarize the conclusions and recommendations of these other conference articles, draw upon the experiences and advice of several knowledgeable parties not involved with the conference per se, and add our own informal opinions. Trying to distill all these points into a coherent plan of action is difficult, but even riskier is an effort to place priorities on the six general categories of recommendations. Yet an attempt is made to cast some order and a time-frame to their development, because of the possible pay-off of a well-thought-out strategy. These are, of course, our own thoughts and others may disagree, but we do feel confident that they provide a reasonable starting point for discussion and action.

SIX CATEGORIES OF CONTROL SYSTEM NEEDS

In reviewing the articles presented earlier in this conference, we compiled a list of the most noteworthy conclusions, observations and recommendations. In our own words and by our own interpretation, those sentiments are contained in Table 1. We have taken the liberty of attributing an author to each of these points so any reader can return to the original source document and review our interpretation.

These observations and recommendations fell quite naturally into six general categories of needs. The mapping between individual points and categories was one-to-one; i.e., no multiple cross-listings were used for reasons of presentation simplicity, though there is acknowledged overlap.

The first and certainly a very important concept category is **Data Connectivity.** Data Connectivity can be viewed as the relating of conceptual or planning-level information (not just data) to the most detailed level. For instance, Nethery points out that preliminary budgets prepared for conceptual designing and feasibility studies have no "tight links" to the control budget which comes from a definitive estimate prepared during detailed design. This is more than just a top-down concept, however. Often, we wish to upgrade factored estimate data-bases with completed project costs (see O'Connor), so a linkage must be available to go from bottom to top.

The issue of **Information Management** brings together the oft-repeated concerns for information availability, security and timeliness. It also spans the data collection stage, though we have separated **Reliable Data Capture** as a distinct function. (Largely because Teicholz, in his very important paper, proposes a lengthy number of data sorts and tests.)

Cost-Schedule Integration has been an identified need for quite some time, yet many obstacles still remain before true achievements are attained. To some extent this could be viewed as a part of Data Connectivity, but we distinguish Cost-Schedule Integration as a literal information connection problem--the same level of cost detail is being tied to (approximately) the same level of schedule detail. This subject is so broad by itself, yet accomplishments seem closer to realization than the top-down-top problems, that a separate grouping was defined.

There are a certain number of recommendations which pertain more to the fundamental ways we practice and pursue project control than to database or computerization issues per se. We have categorized these as **Project Control** issues. These are the strategies, as opposed to implementation or tactical details, that must be addressed. For example, whether we use commitment data in cost reports (as Logcher advocates). A lesser role will, in all likelihood be played by the computer specialists in this area than any of the other five categories.

Finally, **Change Management** may too be thought of as closely related to several of the other groups. Still, we chose to specify it separately since it is a particularly thorny topic, especially when scope changes are involved.

This then is the way we assigned the numerous points of discussion that arose from the preceding papers. Again, Table 1 is the starting point for the subsequent discussion. In the next section we discuss some of the more important recommendations as a way to frame the later discussion of implementation strategy.

Table 1 Recommendations Summary

I. DATA CONNECTIVITY (Conceptual to Detailed level information transfer)

Nethery Preliminary budgets are prepared for conceptual designs; the control budget comes from a definitive estimate prepared during the early part of detailed engineering design.

O'Connor Use the same Work Breakdown Structure (WBS) for conceptual estimating as for later control purposes.

We need closer ties between estimating software and CAD systems to allow "what if" analyses.

Compare early estimates with final costs, not just definitive estimates. Document assumptions that proved false, and assess impacts.

Diekmann Project Control Systems must be designed to handle big and small projects, and should be standardized within a company.

Table 1 (con't)

Separating projects into several smaller jobs is advantageous to project control.

Riggs We need a mapping between the various charts of account that appear on the same project.

Hester Any given cost code is common to several tasks, thus it is difficult to isolate any single task and its costs & changes.

Badger Areas of suggested improvement include: a) better integration with the engineering-design-construction process; b) elimination of errors; c) better visualization of the project with respect construction sequence, equipment access, work accomplished, & problem areas; and d) reduce manpower time required for data input.

II. INFORMATION MANAGEMENT (Incl. reliability & availability)

O'Connor Update historical cost databases with final cost values. Use electronic notebook systems to facilitate such comparisons. Use electronic margin notes for explanations.

Diekmann Systems must be designed with the ability to produce daily information to project personnel.

Hester Need reports quickly and in the field. This will shorten turnaround time, reduce errors and their impact, and encourage more "what if" analyses by the people that count.

Need special fields in data base records for narrative reports by foremen to report unusual conditions; e.g. causes of change.

Badger Stone & Webster performs three separate take-offs for its projects! Tremendous amount of redundancy.

Logcher Computers are taking us away from the feel of engineering.

Existing systems provide data, not information, and they are usually late at that!

Object-oriented programming environments will aid in performance problems.

Exception reporting helps identify problem areas, but it doesn't provide explanations.

III. COST/SCHEDULE INTEGRATION

Nethery There are 2 related yet separate control systems on a project: the Work Breakdown Structure and the Cost Breakdown Structure.

Table 1 (con't)

The financial/accounting/payroll function of the company must adapt to the project controls system, since project control is more detailed. The accounting subsystem should link with the integrated Project Control System so that basic work-hours and cost data are captured just once.

Construction must bow to the needs of start-up, engineering must bow to the needs of procurement & construction, and procurement must bow to the needs of construction. (Actually, more relational, systems-oriented or interdependent attitudes are needed.)

Teicholz An excellent and comprehensive model is presented for linking cost accounts and schedule activities, based on budget and forecast amounts. See his article for a mathematical treatment of cost & schedule integration.

Badger The Engr. Breakdown Structure (EBS) differs from the Construction WBS in that work packages differ in scope.

There is CAD software available for analysis & design that existing Project Control Systems should hook to.

IV. PROJECT CONTROL ISSUES

O'Connor Don't use factored estimates for control purposes unless absolutely necessary. Specify ranges to underscore impreciseness of estimate. Scope is commonly too ill-defined at this time. Better theoretical tools needed.

Conceptual design should include off-sites (utilities, services) as well as on-sites; that is where the extreme cost overruns occur.

Employ risk analysis in estimating & scheduling.

Diekmann About 8% of a design budget is the right amount to spend on design project control.

A hierarchical approach to scheduling is preferred.

CPM is losing its appeal as a scheduling tool.

We need to gain some measure of control over vendor information because of its impact on engr. & design production.

There is a definite trend toward controlling work by work packages. Short interval scheduling is effective.

Riggs The lowest level of a WBS is a task that takes 1 week to complete

Table 1 (con't)

Some contractors are using more summary level WBS's. This lets more control reside in the field.

At the 2 lowest levels of WBS, firms are using bar charts, not CPM.

For companies who regularly use CPM, templates are coming into favor.

Can we integrate CPM at one level with bar charts at another?

Assessing percent complete can be by various methods: Incremental milestones; level of effort; start/finish percent; supervisor opinion; cost ratio; and earned value.

Trend analysis is by several techniques: productivity profiles; performance factors; variance analysis; and progress curves.

Some owners control only key labor & material accounts.

Most forecasting is linear extrapolation.

Badger Performance in an Engineering Management System (EMS) is measured more in man-hours than physical quantities.

Logcher Some systems don't use commitment data, and should.

V. CHANGE MANAGEMENT

O'Connor One owner claims that any change after P & ID's must have a 10:1 payback.

Hester We need systems that allow "tagging" an account with narrative information. This will allow recording of changes, disruptions, etc.

If we had a special field attached to an account, say disturbed vs undisturbed, we could sort those work items that were disrupted by change.

VI. RELIABLE DATA CAPTURE

Teicholz See his article for specific suggestions on important tests to check for.

Hester About 10-20% of the data input for summary reports is miscoded.

DISCUSSION

Data Connectivity

The general idea with this set of recommendations is to provide stronger connection power between conceptual level information and detailed level data. As Nethery notes, preliminary budgets are prepared from conceptual designs, where perhaps only the number of building stories or units of manufactured product are known when a "go/no-go" decision must be made. The control budget comes from a different set of data; namely, a detailed quantity take-off when design is very far along. As Badger observes, this normally means entirely separate take-offs, which is expensive and error-prone. By tying these different information levels together, O'Connor's wish for updating early estimate historical data by tracking against actual, final costs would be realizable. This idea would also ease the development of CAD/CAC (Computer-Aided Design/Computer-Aided Construction) systems that, for example, allow graphical display of construction erection sequences from original design drawings.

Hester identifies the practice of using the same cost code to record performance data for activities which may be separated for scheduling purposes; e.g., reinforcing steel for the east wall versus the west wall. A rational database architecture scheme may aid such separate tabulations.

Diekmann advocates that firms use a standardized control system for most projects, big or small. There is lots of room for skepticism on this point, but the advantages would certainly be legion. Riggs, in effect, is supporting a move in that direction when he comments on the growing use of schedule and cost system templates. The ultimate, of course, would be to shift to one generalized, generic template that fits most of a contractor's project types.

Levine (1986), in a very good textbook on the subject, criticizes existing control system technology on two fronts. First, systems generally do not allow allocation of fixed costs across an activity's duration in any non-uniform fashion. Yet when we schedule re-steel installation for a floor slab, one day's worth of work may be expended to shake-out four day's of actual productive work. His other major complaint is that we still draw networks by hand, and maybe even do some alterations by hand, before we enter the data into the system. (Some people still use data coding sheets!) It would certainly be pleasant to have an easy-to-use electronic network graphing system.

Information Management

The issues surrounding information management can be regarded as information capture, manipulation and reporting, including availability and accessibility. (Reliability and error handling are dealt with separately.) To an extent data architectures are the subject; e.g., Hester's call for special fields and records and Logcher's optimism for object-oriented programming structure.

Two other points are especially noteworthy here, the first being O'Connor's suggested use of an electronic margin note concept when updating an estimating database. Virtually everyone agrees that cost records should be kept current by using recent, historical experience. The problem arises when a user would like to incorporate some new data into the database, but finds it to only be similar, not identical in nature to an existing account. A margin note system feature would be valuable to record these special exceptions.

The other important point is Logcher's comment that exception-based management reports are useful in identifying the accounts with large variations, but they don't explain why the variation occurred. (The difference between data and information.) This is, of course, the root of management control--the "why question." The electronic margin note idea may be useful here as well, but likely won't be a complete solution. Variances usually occur for a multiplicity of reasons, many of which may be beyond the scope of view of the person doing the reporting. What can't be seen or understood is difficult to explain. In all likelihood a better understanding of the relational, not just hierarchical, mapping between project accounts will encourage more complete understanding.

Cost/Schedule Integration

People close to the industry have been clamoring for better cost-schedule integration systems for years. Some professionals believe that a control system combining both information types will allow more complete understanding of field problems and implementation of effective solutions. The problem is that schedule work breakdown structure is defined much differently than the cost breakdown structure, as Nethery notes. We schedule the construction of an east wall separately from the west wall, normally, but aggregate their costs. Teicholz does offer an interesting, flexible model approach. It requires effort to maintain parity between the schedule breakdown system and cost breakdown system when actual costs vary significantly from planned.

Another issue that Badger raises, being a turnkey contractor, is the relation between the construction and engineering work breakdown structures. This is integration in a much broader context--reminiscent of the data connectivity issue raised before. Tackling this big issue will eventually allow the integration of systems between the CAD and construction phases.

Project Control

The subject of project control pertains to the concepts, procedures and techniques used to measure, analyze and report performance. In this category are some of the most important yet difficult recommendations to implement. O'Connor, for instance, faults factored estimating techniques as presently practiced because of historically poor reliability. His recommendation of specifying a cost estimate range, rather than single point value, is reasonable and comparatively easy to implement. Statistical concepts, e.g., probability distributions, are appropriate here. Such risk analysis techniques have been available in the research literature for some time (e.g., probabilistic estimating, PERT), though notable implementation issues remain.

Diekmann's observation that CPM has lost appeal as a schedule control tool seems to argue against more sophisticated, complex techniques, like risk analysis techniques. It seems that simplification of these types of tools, combined with increased familiarity and comfort as attained through continued education, may be the keys.

Riggs stresses two important points, the first being that greater connectivity between bar charts (at the lowest level of project control) and a network representation (at higher levels) is desired. One-to-one mapping and logic constraints are difficult to preserve when moving between different schedule control techniques. His second point is that more performance forecasting is subjective, and when more analytical, it is almost always linear extrapolation. Many people have tried to apply more advanced nonlinear models, with a modicum of success. The whole area of learning curve theory, moderately successful in the aircraft industry, still remains to be applied widely in construction. Usually, the discrete tasks are too short in duration to allow the phenomena to establish itself. Riggs' supplemental point of using more summary level task definition (minimum task duration of one week) could well be a step in the right direction.

A final issue in this area is Logcher's advocacy of using commitment data. Some planners embrace this idea. A substantial number argue that such data are of more financial concern than project concern, thus clouding the picture, and should therefore be excluded. The remedy here is to allow system user choice.

Change Management

Like cost-schedule integration, the subject of change management seems to have haunted project managers for years. Hester's support of existing data base management technology is, of course, quite pragmatic. Building "hooks" between a project control system and software, like dBase III+ or R:Base 5000 (TM) could and should be made easier. The most critical problem in this broad area of change management still defies an easy-to-use and universal solution: How to measure impact or ripple effects. Research is known to be underway in the Construction Industry Institute and elsewhere that may lead somewhere.

O'Connor's point about one owner demanding an extremely high pay-off threshold for any proposed change is a reaction to the contentiousness surrounding change impact measurement.

Reliable Data Capture

Spotting and correcting input errors before the data goes too far in the control system has been nettlesome, whether the system is computerized or manual. A pre-processor with a variety of variance tests and sorts, as discussed by Teicholz, is a sound quality control program for data management. Still, there is no substitute for doing it right the first time, and that requires commitment to project control by all the people in the organization, especially the crew foremen who report the time and perhaps quantity data. In return for that commitment to accuracy, those data collectors should see the value of their efforts. Such pre-processing may also eliminate some of the data "game playing" that inevitably occurs by forcing proper categorization.

A PROPOSED IMPLEMENTATION STRATEGY

Just as important as what needs to be done, is the order and time-frame in which these enhancements should be brought on-line. With the customary warning that these are the authors' ideas, the following implementation strategy is suggested.

Any long-range development plan needs to have intermediate check-points. The implementation strategy suggested here and represented by Table 2, has such milestones. The recommendations and observations noted by the other authors of this Specialty Conference can be categorized in two ways: by category type and by probable time required to achieve the specific goal. Thus Table 2 contains near-term, intermediate-term and long-range goals. The key here is to build a framework that not only allows solution of individual problems, but also leads to some comprehensive, integrated system.

Without doubt, the Industry is headed toward greater unity of information across all the current boundaries of project development;ment, planning, design and construction. The major impediment to realizing this goal of project integration is the fragmentation or lack of uniformity in practice between companies. As Riggs has described, there is no one standard way of forecasting future labor costs, or discussing productivity, or even measuring project performance. There are segments of the Construction Industry large enough to justify development of project control systems specially tailored to their needs, just as there are cost estimating systems available today geared strictly for the building or mechanical segments. Moreover, there are major areas of project control practice that are common to the whole Industry, such as bar chart scheduling. These two points, large industry segments and some degree of common practice, are encouraging enough to stimulate further project control systems research and development. Though the plan in Table 2 may not be realized as exactly prescribed, this suggested implementation strategy should provoke discussion and intelligent decision-making about which direction(s) to pursue and at what pace.

Table 2 Steps Toward Better Project Control

Recommendation Category	TIME-FRAME Short-term	Intermediate-Term	Long-term
I. Data Connectivity	Common WBS from concept to detailed planning; updating conceptual cost data with actual costs	Schedule templates; linkages between estimating systems and CAD software	Graphical representation of project erection; reliable mappings between WBS's for cost and schedule
II. Information Management	Quicker report turn-around; special narrative fields tied to WBS accounts	Exception reporting with explanations; electronic margin notes; minimization of quantity take-off	Object-oriented data representation; systems that provide information, not just data
III. Cost/Schedule Integration	Teicholz's recommended treatment for allocating cost and schedule data	Integration of cost and schedule work breakdown of structures	Integration of engineering and construction work breakdown structure
IV. Project Control	Develop user options for commitment data use	Improve forecasting	"Intelligent" systems that aid selection of key accounts for performance assessment; improve communication with vendors and suppliers
V. Change Management	Integrate existing data base managers with existing project control systems	Develop better tools for change impact measurement including accurate assessment of ripple costs	Develop better planning tools to identify and predict change impact as a "what if" tool
VI. Reliable Data Capture	Develop better preprocessors	Provide field personnel with motivation for "getting it right the first time"	"Intelligent" error detection and correction; e.g., diagnostic expert systems

SUMMARY AND CONCLUSIONS

The objectives of this ASCE Specialty Conference were to track the current state-of-the-practice, identify areas in need of refinement or thorough re-thinking, and generate possible solutions to these problems. The whole range of project control has been discussed here: from design to construction, and from estimating to scheduling to cost control.

In this paper the principal recommendations and observations emerging from the preceding papers have been summarized into tabular form, Table 1. Six categories of particular concern were identified and individual recommendations grouped under one such category heading. Relationships between these various recommendations, both within and between the six category headings, were identified and discussed. The other major contribution of this paper then was to mold these needs and solutions into a proposed implementation strategy; Table 2.

With the usual caveats about such personal opining, the authors suggest this as a formula for the industry, practitioners, system developers and researchers as a whole to pursue. These and many other new and innovative technologies and concepts that were not addressed in this conference because of time limitations also await exploitation. Include here are subjects like natural language processing, voice and optical pattern recognition, and automated data sensing.

Taken together then, the future certainly seems brighter than ever before for effective and easy project control. The task before us is to experiment with the various recommendations presented at this specialty conference and refine them to a state of practical usefulness. The time to begin is now.

REFERENCES

Levine, H. A. (1986), Project Management Using Microcomputers. Osborne McGraw-Hill, Berkeley CA.

SUBJECT INDEX

Page number refers to first page of paper.

AUTHOR INDEX

Page number refers to first page of paper.